C. 2

574.1
    Still, Henry
       Of time, tides, and
    inner clocks

DATE DUE

MAY 2 8 1985			
JUL 1 7 1985			
MAY 1 8 1999			
JUN 1 5 1999			
JUL 1 3 1999			
AUG 2 0 1999			
SEP 1 8 1999			
JAN 0 5 2000			
FEB 0 2 2000			

OF TIME, TIDES, and INNER CLOCKS

taking advantage of the
NATURAL RHYTHMS OF LIFE

OF TIME, TIDES, and INNER CLOCKS

by **Henry Still**

:red keRneR/publishing pRojects

n association with

Stackpole Books

OF TIME, TIDES, AND INNER CLOCKS
Copyright © 1972 by
Henry Still

Published by
STACKPOLE BOOKS
Cameron and Kelker Streets
Harrisburg, Pa. 17105

Printed in U.S.A.

Library of Congress Cataloging in Publication Data

Still, Henry.
    Of time, tides, and inner clocks ... taking advantage of the natural rhythms of life.

    Bibliography: p.
    1. Biology--Periodicity. I. Title. [DNLM: 1. Biological Clocks. QT 167 S857o 1972]
QP84.S77   574.1   72-6094
ISBN 0-8117-1140-4

There is a season for everything
And a time for every purpose under heaven;
A time to be born and a time to die;
A time to plant and a time to reap. . . .
                            —Ecclesiastes

Contents

CONTENTS

Introduction

MODERN MAN LIVES by the clock but, very often, it is the wrong clock. He times his hours of sleep, work, and play by the mechanical clock he has invented, but the functioning of his body—indeed, the functioning of all organisms—is timed by a biological clock, or clocks, which regulates sleeping, breathing, living, and dying in apparent synchronization with the universe. Most people cannot hear or obey the ticking of their inner clocks because their lives are run by the artificial clocks of modern society.

It is only recently that science has begun to show interest in these biological clocks. A century ago scientists began to see manifestations of mysterious internal machinery in the cycles of life. Flowers and some plants opened by day and closed at night. Birds knew unerringly when and where to migrate. Some animals knew when and how long to hibernate. Later, observations of natural changes in plants, animals, and men were moved into the laboratory and tested statistically, with and without the environmental factors which had been blamed for triggering the behavioral patterns of living organisms.

As a result of these studies, according to Dr. Bertram S. Brown, director of the National Institute of Mental Health, "Many aspects of human variability—in symptoms of illness, in response to medical treatment, in learning and job performance—are being illuminated. Already some of our changes of mood and vulnerabilities to stress and illness, our peaks of strength and productivity, can be anticipated. Moreover, by the end of this decade, much that is still considered unpredictable in health and human performance may become foreseeable through research into the nature of biological time cycles." [1]

How will this new knowledge benefit man? There are several intriguing possibilities. Perhaps employers will allow their employees to work during the hours of their greatest efficiency, thereby staggering work hours and incidentally eliminating the discomforts of rush hour traffic. Doctors may prescribe that drugs be taken at those times when they will be most effective. Surgeons may perform operations when their patients are at the peak of their physical vitality. People may be taught to cure various ailments by altering their own blood pressure and temperature.

But all the benefits of this new science of biochronology do not lie in the future. Many practical applications have already been made. Ways have been found to minimize the disruption of body rhythms caused by changing work shifts or flying across several time zones. Some insomniacs have helped themselves to sleep better by changing their hour of retiring to a time when their bodies were ready for sleep. Researchers have found out much about scheduling meals for those times at which the human body can best absorb nutrients.

As man learns to live in harmony with his inner clock, as he learns that there is indeed "a season for everything and a time for every purpose under heaven," he will perhaps achieve some of the serenity of what Pythagoras called the "harmony of the spheres" and maybe even discover that the classical Greeks were right when they ascribed an influence on human affairs to the heavenly bodies.

CHAPTER 1

Are we marked by the stars at birth?
A hard look at the claims and credibility
of astrology—what truths lie at its core,
what are the cosmic influences on human life?

The Harmony
of the Spheres:
More Than Fantasy?...

MODERN MAN, INSTRUCTED by science that there must be a physical cause for every effect, pushes astrology to the back of his mind along with alchemy and the black arts of magic. Yet scientists are haunted by the reminder that astrology has been with us 5000 years, whereas the period of modern insight into the physical nature of the universe is little more than two centuries old.

By scientific hindsight it is easy to see the foolishness in old superstitions. Yet men have always known no more than they could learn by observation. In their ignorance and vulnerability to natural enemies, the first men assigned good and bad traits to the forces of nature which seemed to control their lives completely.

. . . The Development of Astrology

THE sun became a god and the moon a goddess because the first men could see that these creatures of the sky ruled day and night. Without knowledge of their nature, it would have been simple logic to assign these heavenly bodies specific powers related to the birth, activities, and death of men. The seasons passed and returned just as the constellations wheeled overhead, each dominating a spring or winter, summer or fall, then going to rest while new stars came to govern the sky. The years, too, found their recurrent patterns in the wanderers, the planets, which seemed not to follow the rhythmic order of sun, moon, and stars until the first wise men learned to measure their longer cycles of turn and return. Then the priest-astrologer tried to use his proven knowledge of the skies to foretell the future of kings and world affairs.

Masters of early astrology were the Chaldeans and Sumerians, who studied the sun, moon, planets, and stars more than two thousand years before Christ. From rhythms of fertility and natural disasters, coinciding with certain positions of planets and constellations, the priest-astronomers drew a system of apparent cause and effect. Their followers must have been impressed when an eclipse of sun or moon occurred exactly when the priests predicted it. Backed by such proof of their power, astrologers were encouraged to extend their predictions more boldly to the cyclic affairs of men.

By the time Chaldea was conquered by the Greek warriors

of Alexander the Great in 331 B.C., the pretensions of astrology extended to the prediction of individual futures. The Greeks had recognized the five (then known) wandering planets 600 years before, but now they adopted the astrology of the Chaldeans and refined it by philosophy and mathematics to the form in which we see it today.

Two hundred years before the Alexandrian conquest Pythagoras, who believed the universe was a sphere containing the earth and the air around it, gave us the mystical "harmony of the spheres." He saw the heavenly bodies revolving in concentric circles around the earth, each fastened to a sphere or shell around the earthly universe. The swift revolution of each of these bodies, in the master's poetic vision, caused a swish or musical hum in the air. The music of each planet was in different pitch or rhythm—just as the string on a musical instrument vibrates fast or slow according to its length. Thus Pythagoras saw the planetary orbits as a huge lyre whose strings were curved into circles.

Plato saw the sun and stars, not as celestial bodies, but as gods. Aristotle also defended the notion of the divinity of stars, stating: "This world is inescapably linked to the motions of the world above. All power in this world is ruled by these motions." [1] The classical Greeks devised such complex and logical geometries with the stars and their shifting positions that the pseudoscience of astrology has been impossible to dislodge completely from the rational mind.

Later astronomers ungallantly removed the earth from the center of the Pythagorean universe. The harmony of the spheres, however, still echoes in our souls as an instinctive recognition that life on earth vibrates in tune to the passage of sun, moon, and planets, the earth's revolution, and of stars so far away that the light we see was shed before the Chaldean civilization was born.

. . . Unknown Life-regulating Forces of the Universe

Astrology has persisted stubbornly through the ages probably because scientists have not yet proved that living creatures are *not* influenced in their cyclic variations by subtle forces emanating from the planets, sun, moon, and the earth itself. Every living thing responds to the rhythm of daylight and dark, changing seasons, vagaries of weather, and the longer cycle of years which leads to death. But as centuries have passed with their growing scientific observation and synthesis of ideas, it has become apparent that plants, animals, and men also behave according to more subtle rhythms. They are equipped with internal "senses" which respond to forces not yet measured, and which continue to operate when the apparent triggering outside force is blocked off.

What are these as yet unmeasured forces? Science has no firm answers to this question at present, but there is evidence that some of these forces may emanate from the planets, as astrologers have always contended. Some modern scientific findings suggest that the ancient astrologers came to a dead end in developing their "science" because they did not know how to ask the right questions or did not have tools to establish a statistical base of proof for their hypotheses. Thus it may be only the frozen rules, the so-called laws of astrology, that are wrong.

During recent years investigators of the periodic clocks which govern the life cycles of plants, animals, and men have reopened the blocked matrices of astrology and replaced it with a new science. Subtle forces from the cosmos apparently *do* influence our lives by physical laws not yet fully defined.

Evidence of this is found in the surprising results obtained in a study by Michel Gauquelin, noted investigator of biological rhythms, and his associates in France. While preparing a critique of traditional astrology which completely demolished the "laws" of that pseudoscience, the investigators compiled research samples of the birth dates of 576 members of the French Academy of Medicine. Although the facts did not fit any traditional rule of astrology, the researchers found that a statistically large number of great physicians had been born when the planets Mars and Saturn had just risen or culminated in the sky. With scientific skepticism, Gauquelin compared the findings with a sampling of average people but found no more than random grouping of profession or skill under any planetary sign.

Not satisfied with the original statistics, the scientists repeated the study with 508 other physicians. This time they established not only the date but also the precise hours of birth. Again the findings revealed the unusual grouping with more than an expected average of doctors related to Mars and Saturn. Because this result was embarrassingly similar to astrological foreordination, the French investigators set out to collect a more massive sampling. Traveling through Europe they accumulated more than 25,000 birthdates, including doctors, writers, actors, politicians, athletes, military men, and others of prominence.

As they computed the planetary positions at the moment of birth, the strange pattern persisted. A more and more precise statistical relationship appeared between time of birth and professional career and, even more strangely, the data showed that each professional group seemed to have a planetary clock of its own.

A great many persons born when Mars was appearing over the horizon or at zenith in the sky later became famous doctors, athletes, or military leaders. Future artists, painters, or musicians seldom were born under the same sign. Actors and politicians were born more frequently when Jupiter rose or reached zenith, but scientists rarely were born at that time. Thus, as far as vocational success was concerned, the moon, Mars, Jupiter, and Saturn were found to act as planetary clocks. A cosmic pulsation seemed to cause, during the 24-hour cycle of the day, more births of future doctors at certain times, more future artists at others, and so on.

Like a hunter who catches the scent of rare but elusive prey, Gauquelin sought a hypothesis to fit the statistics. One possibility, which seemed too tenuous to support, is that some form of radiation from the planets marks a baby at birth with an influence which persists through life.

Another possibility is that a child inherits from his parents the tendency to be born when a planet rises or is at zenith in the same way he inherits the color of his hair or shape of nose. To test the theory of planetary heredity, the investigators moved on to find if there were similarities between the position of the planets at the birth of parents *and* their children. Five years' work with birth records of 30,000 parents and children were subjected to statistical analysis. Results showed 499,999 chances to one that planetary heredity is real.[2]

... Scientific Theories About Biological Clocks

IT requires more than one tree to make a forest, but Gauquelin's findings go far to support one of two major theories which attempt to explain the internal biological clocks which govern the periodic rhythms of all living organisms. The first postulates roughly that the cyclic governors are inherited and through countless ages of evolution became so engrained that they are independent (*endogenous*) of external cues. The second theory states that the internal timers are special sensors (or extensions of our known senses) which coordinate our functioning when triggered by outside (*exogenous*) forces— light and dark, changing seasons, gravity, and electromagnetic waves from the earth, sun, moon, and possibly the planets.

A foremost proponent of the second theory is Frank A.

Brown, Jr., professor of biology at Northwestern University. He and his coworkers have been studying plant and animal rhythms for more than thirty years and have become convinced that there are tides in forces from outer space and that these are the master clocks which synchronize our living timekeepers.

These cosmic oscillations, according to Brown, are bound to phases of the moon, sunspot cycles, and perhaps radiation and gravitational force from the planets. Such gentle tides, though not yet measured by scientific instruments, are believed to ebb and flow in hourly, daily, monthly, and even yearly rhythms and drive other geophysical forces which surround the earth. These may include barometric pressure, earth's magnetic field, subtle changes in gravity, atmospheric ionization, cosmic rays, and weak electromagnetic fields.[3]

In subsequent chapters, we will see that the tides of life exist in real and measurable form. Chapter 2 examines evidence that certain sea animals such as fiddler crabs and oysters time their periods of feeding and rest according to the gravitational pull of the moon. The reproductive cycles of other ocean organisms—including grunions, the fireworms of the West Indies, the Palolo worms of the Southwest Pacific, and certain seaweeds—are governed by the phases of the moon. But of greater interest to man is the fact that there have been recorded cases of hemorrhage, epilepsy, and other ailments among humans which seemed to be closely connected with phases of the moon. There is even evidence that sunspots may affect human behavior, since a correlation has been found between their occurrence and emotional outbursts among mental patients.

Admittedly, science does not yet know much about the mechanisms by which the sun and moon regulate the behavior of living organisms. Much more is known about the ways in which light itself—whether from the sun, moon, or electric lightbulbs—affects the behavior of life forms, from plants and insects to man (see Chapters 2 and 3). The hours of light and dark determine when plants open their leaves, when fruit flies emerge from their pupae, and when bees gather nectar. In mammals, including man, light affects the pineal gland, which most scientists believe to be a coupling device regulating the phase relations among biological rhythms. There are indications that light, or the lack of it, may stimulate the production of adrenal and reproductive hormones in human beings.

Whatever may be the environmental forces which regulate the biological rhythms of living things, man shares those rhythms with other forms of life. The study of various forms of life may shed light, therefore, on those physiological clocks in man which affect his behavior so profoundly.

> *Persistence of the ancient cult of astrology into modern times has moved scientists to measure the subtle tides of change in natural and human affairs. Although there appears to be no scientific foundation for the traditional practice of astrology, modern research indicates that all life operates by internal rhythms keyed to the motion of sun, moon, stars, and cosmic forces not yet clearly defined. Since these rhythms are found in man as well as in all other creatures, a study of periodicity in lower forms of life may help to explain how the biological clocks of the human body operate.*

CHAPTER 2

The mysteries of built-in timetables and compasses. What laboratory investigations reveal about these mysterious synchronizations and the setting and resetting of nature's clocks.

Nature's Feats of Split-Second Timing . . .

IT IS A damp, cool summer night on a southern California beach. The moon, a day past full, has topped low mountains to the east, shining eerily through rolls of mist hanging above long breakers of surf rushing up the sand.

Hundreds of people line the beach, equipped with pails, pans, jars—even T-shirts serving as impromptu bags. Children run excitedly up and down the wet smooth sand, playing hide-and-seek with the waves breaking white in the moonlight. Others wait patiently.

Tonight the grunion are running.

Many have never heard of the grunion. Others—particularly young girls convinced their boy friends cooked up the story to lure them to the beach on a moonlit summer night—doubt they exist. But the grunion, elusive as they may be, are very real.

Now there is a subtle change in the misty night. Only the initiate would know the tide has reached full and is turning. A few minutes more and even the children are quiet now, waiting. Suddenly a new wave breaks across the sand, its white froth alive with thousands of tiny silvery fish about twice the size of sardines. The grunion are in!

As the wave recedes, the wet sand is covered with squirming fish. Human predators pounce, scooping up the wriggling prey with their hands (the only legal way) and dumping them into waiting receptacles. Another wave breaks with more grunion. No one minds being drenched by the cold Pacific water in the shrieking, laughing excitement of snaring the slippery little fish which sparkle like sequins on the moon-drenched beach.

An hour later the ocean returns to normal and the fish are gone. Many take their catch and go home. Young people cook their feast over driftwood fires, then linger into the night pondering this strange ritual of fertility.

This is the story of how the grunion reproduce.

There is no promise they will come ashore at every beach, but the fish may be found from Ensenada in Baja California to Santa Barbara several hundred miles north. They time their spawning runs at two-week intervals, a day or two after the moon is full or completely dark. In other words, they come ashore after the highest tides of the month have passed, and then time themselves to appear almost exactly fifteen minutes after the tide has crested that particular night.

As the thousands of fish surge ashore, the female dances upright on her tail and digs her back half into the sand. She deposits her eggs in the hole. The male wriggles up and curves his body to wrap around her as she sways. In a few seconds before the next wave comes, he has fertilized the eggs and squirms back toward the water. The female also works free of the hole she has dug, flips toward the water, and vanishes into the breakers. Wet sand slides into the hole and covers the eggs. They have thus been given the best possible chance to hatch new fingerlings.

The high tides of the month and the specific night both have passed so the eggs will not be washed out of their hiding place prematurely. Two weeks later they will hatch exactly in time for the new high tide to wash the tiny fish out to sea and join their parents. This depends, of course, on the grunion's ability to outsmart human predators who have learned their cycle.

The love dance of the grunion is itself a marvelous thing, but even more mystifying is the internal clock which enables these fish to choose the exact night and minute to lay their eggs. Do they sense the changing moonlight? Are there subtle gravitational changes which signal exactly when to come ashore, or is the "knowledge" engraved by evolution somewhere in their brains or nervous system?

No one knows for sure, just as no one knows exactly how or why the salmon finds its way back to spawn at a certain time in the exact freshwater stream where it was born. The grunion and salmon are two of the more spectacular examples of how all living organisms appear internally geared to function in orderly cycles. No two species seem to respond in exactly the same way to known influences such as light and dark, temperature change, or the tides.

Most life cycles seem to fluctuate with the 24-hour period of day and night in the earth's rotation. These are known as *circadian* (from the Latin for "about a day") rhythms. Processes which operate in shorter cycles, down to seconds or microseconds in basic cell processes, are termed *ultradian*. Longer rhythms, such as mood changes occurring at regular intervals of days or illnesses which recur at certain times of year, are *infradian*.

Of all the cycles in life, the circadian rhythm of organic change from day to night and back again in 24-hour periodicity is most pervasive in living things and was the first such rhythm to attract scientific notice. "That period of 24 hours, formed by the regular revolution of our earth, in which all its inhabitants partake, is particularly distinguished in the physical economy of man," wrote C. W. Hufeland in 1797. "It is, as it were, the unity of our natural chronology." [1]

. . . Living Plant Clocks

LONG before that, however, a note about circadian rhythms was passed down to us from the days of Alexander the Great three centuries before Christ. Androsthenes, who marched with Alexander on his conquests of the East, noted while marching through India that the tamarind trees opened their leaves during the day and closed them at night.[2] It might seem commonplace now to observe a plant responding in such fashion to the change from light to darkness, but in ancient days the delicate shifting of a pattern of leaves was more often obscured by such grosser happenings as the birth of kings and the clash of battle.

After Androsthenes twenty centuries passed before the French astronomer Jean Jacques d'Ortous de Mairan discovered that plants can tell time even if the sun is absent. It was 1729, in Paris, and de Mairan, who bent his mind to many studies and speculations, absent-mindedly experimented with the heliotrope, a plant so named because it turns its face to find the sun. He learned, by keeping the plant in darkness, that it continued to open during the day and close during the night. This was the first controlled experiment in testing the biological clocks of lower life, but most contemporary scientists ignored it.

Thirty years later the experiment was repeated by Henri-Louis Duhamel. He chose a wine cave as a place which surely would be free of all changing environmental triggers such as light, temperature, and humidity. While carrying a plant into the cave, Duhamel accidentally jostled it. The plant abruptly closed its leaves, but he left it in the cave, certain that his clumsiness had ruined the experiment. When he checked it at 10 o'clock the next morning, he found the leaves were open in complete darkness. Moreover, the plant continued its sleep-waking cycle in the dark cave for many days. Late one day, while its leaves were still open, Duhamel carried it outside. It remained "awake" all of the following night.[3]

Thus it was demonstrated that at least some plants contain an internal clock which causes them to function in their normal manner even when known environmental influences are blocked off. Beyond that there were indications that the clock could be "adjusted" to follow different cycles.

In the nineteenth century, the Swedish naturalist Carolus Linnaeus was first to notice that various flowers opened at different hours during the day. The charmingly practical application in formal gardens of Europe were flowers planted to form a clockface with varieties in each bed blooming and closing at different hours.

One common arrangement started with the spotted cat's-ear, which opens at 6 A.M., followed by the African marigold at seven and the hawkweed at eight. Nine o'clock was marked by the closing sow thistle, and the nipplewort closed at ten. At 11 A.M. the star-of-Bethlehem opened, followed by the passionflower at noon. The afternoon was marked by the time of closing of other flowers: childing pink at one, scarlet pimpernel at two, hawksbit at three, bindweed at four, and the white water lily at five. The evening primrose opened at 6 P.M., completing the 12-hour day.

Many observations have shown internal periodicity at work in insects, animals, and sea creatures as well as plants, although it is only in recent times that biological timing mechanisms have been deemed worthy of serious investigation. As we look more closely, we find that precise rhythmicity is a basic key to many vital processes.

Artificial conditioning causes us to think of a time-measuring device in terms of springs, gears, and a windup key. Such conditioning leads us, perhaps unconsciously, to search for a clot of cells or nest of nerves in living tissue corresponding to our limited understanding of the passage of time. The rhythmic core of living organisms may be more analogous to the electric clock, which simply oscillates to a measured beat according to electric signals which it receives from the outside world. Such a concept may help in understanding the grunion and other creatures which behave peculiarly in response to signals others cannot perceive.

. . . Sea Creatures That Live by a Lunar Clock

OTHER sea dwellers which exhibit precise timing in their sexual behavior include the fireworms of the West Indian reefs. As early as the 1930s the marine biologist from England, L. R. Crawshay, noted that this marine worm in the Bahamas timed its nocturnal sex activities by the phases of the moon. Groups of females would appear suddenly at the top of the water, shedding eggs and discharging streams of brightly luminous secretion. Then the males would flash like fireflies and dart in to fertilize the eggs. Within a few minutes the ritual was done, but it always occurred just one hour before moonrise on the night before the fourth quarter of the moon.[4]

In the Southwest Pacific, the Palolo worms swarm in huge numbers on the nights of the third quarters of the October and November moons, liberating their reproductive elements into the sea just as dawn breaks. So precise is their schedule that island natives are waiting when the 18-inch worms come seething to the surface. As American midwesterners gather wild mushrooms in the spring, the islanders harvest the breeding worms as a delicacy. Cakes made of the gelatinous creatures are fried and taste like oysters.

Reproductive rhythms also occur in certain seaweeds. Dictyota, a form of brown alga, produces eggs and sperm on a monthly schedule. All plants in one area may be synchronized with one phase of the moon, while those in another area will be regulated by another phase, determined in some manner by the local tides.[5]

Frank Brown at Northwestern University has conducted extensive studies of marine organisms, correlating their activities with lunar cycles and external influences. He started with the premise that all chemical reactions (the basis for all metabolic processes) occur rhythmically and vary according to temperature. Metabolic actions increase in speed when temperature rises, and slow down when temperature drops. Any reaction opposing this trend could be interpreted as responding to some other set of laws.

Brown chose the fiddler crab, which turns dark during the day and pale at night in precise 24-hour circadian cycles. Crabs first were placed in continuous darkness at steady temperature. They continued to turn dark and light on exact schedule though removed from the tide and influence of sun and darkness. Brown then increased the temperature of the water in which the crabs were kept from 61° to 79° but, in test after test, the rhythm of skin color change remained exactly as it had before.

Researchers also discovered the fiddler crab contains another biological clock. Although its skin changes according to the sun, its feeding habits are geared to the moon. At low tide it runs down the beach to scour for food; at high tide it scuttles up the beach to rest. Tides are timed to the lunar day of twenty-four hours, fifty minutes. Brown found the crabs held to this schedule even under laboratory conditions, indicating their clocks operate independently of outside signals, or respond to lunar forces more tenuous than the ocean tide itself. Brown favors the latter.[6]

Other tests centered upon oysters, which open their shells widest at the time of high tide. In one experiment, a batch of oysters collected from the Atlantic Ocean was shipped to Evanston, Illinois, and kept there in pans of sea water in a dark room. For a time, although a thousand miles from their native sea and tides, the oysters continued to open and close according to the old time schedule. By the end of two weeks, however, they had reset their clocks. Now they opened and closed in a new rhythm keyed to high and low tides which would exist if the ocean had moved to Evanston.

The oysters apparently responded to times when the moon was directly overhead or on the opposite side of the earth, when maximum gravitational pull produces twice-daily high tides. The sea creatures may contain a cellular mechanism which responds to the moon's gravitational pull or to minute changes in air pressure caused by atmospheric tides.[7]

Since life emerged from the sea, there is an instinctive tendency to believe in the turning earth and tug of sun and moon as forces bearing directly upon the rhythms of sea organisms as well as our own. It perhaps awaits only the development of adequate scientific techniques and tools to measure and define what other forces are at work. Other creatures, not bound to the sea, offer more evidence that internal biological clocks are functioning. One is the honey bee.

.. Experiments in Resetting Insect Clocks

NEARLY a century ago, a Swiss naturalist, August Forel, observed that bees visiting his breakfast table each morning in search of food always arrived at a regular time, even when no food was available. He concluded "that the bees remember the hours at which they had usually found sweets . . . and that they had a memory for time." His conclusion was born out about 1950 by Karl von Frisch of Austria.

Von Frisch captured a number of bees, marked them, and then offered sugar water at an artificial feeding place from 9 to 10 A.M. each day. The bees quickly trained themselves to appear at feeding time, but this did not determine if they operated by internal timing or environmental signal. To answer part of that question, researchers kept colonies of bees in a salt mine 600 feet deep in the earth with constant light, temperature, and humidity. In all cases the bees remained punctual.

Tests to prove or disprove if the bee time-clocks were independent of external events came in 1966. Insects were trained in Paris to collect food from 8:15 to 10:15 P.M. French Daylight Time. They were then flown to New York and tested again. There they came to feed at 3 P.M. EDT, approximately twenty-four hours after their last feeding. None came at 8:15 to 10:15 New York time. The experiment was repeated from New York to Paris with comparable results.

The question still remained whether sun-altitude or changing daylight would alter the bees' natural timing. In 1958 a number were trained in Long Island to collect food at a certain hour, then flown to Davis, California, with a difference in local time of about three hours, thirty minutes. For a time the bees came to feed exactly twenty-four hours after their original Long Island time. Then one day their timing shifted by one and one-half hours. An equal shift occurred the second day, and on the third they adjusted another thirty minutes. Thus the bees gradually reset their feeding clocks to a local time schedule. Out of these tests came conclusions that bees have true internal clocks which run on circadian rhythm and can be reset or adjusted to new 24-hour cycles.[8]

Similar clocks are evidenced by other insects, including the common fruit fly, which has served science faithfully in other areas of study (among these, valuable experiments in genetics). Here again, however, the experimenters are handicapped by the fact that they can analyze internal clocks only by observing the rhythmic phenomena they time. This would be similar to the difficulty of analyzing an internal combustion engine by watching a car drive down the street.

In the natural world, the adult fruit fly emerges from its pupal case around dawn after seventeen days of development. In the laboratory, if eggs are laid and develop in continual darkness, the young flies emerge at random times of the day. If the maggot larvae are subjected to even a single flash of light during this period of darkness, however, the flies, when they emerge, will come out of their pupal cases at the precise time of day they were exposed to the light flash. Thus, under normal conditions, the insect's timer is set to local sun time while it is still a maggot, and the clock alerts the fly to emerge at dawn. If the dawn is missing, any light source apparently will do. The fruit fly thus has an internal 24-hour clock which seemingly must be activated by an external stimulus.[9]

. . . Wonders of Migration and Hibernation

ONE of the traditional wonders of nature is the migration of birds and how they do it. Their decision to move north or south is geared to changing seasons and probably to the hours and angle of sunlight, but their navigational methods are even more remarkable. As with the swallows which return the same day every spring to San Juan Capistrano in California, birds and their young often find their way back to ancestral nesting sites.

Gustav Kramer, born in 1910 in Germany, was intrigued by the arctic tern, which nests within 100 miles of the North Pole. In the fall it flies over Canada, across the Atlantic, down the west coast of Africa and around the Cape of Good Hope to winter feeding grounds below Port Elizabeth.

The New Zealand bronzed cuckoo flies 1200 miles across the Tasman Sea to Australia and from there 1000 miles north over the Coral Sea to wintering areas in the Bismarck Archipelago and the Solomons. Also the cuckoo young, which have never traveled the way before, do so unerringly without their parents, which have gone on a month ahead.

After World War II, Kramer found part of the answer to the pinpoint accuracy of bird navigation. At the Max Planck Institute in Wilhelmshaven, Germany, he contemplated the fact that caged birds exhibit what has been called "flight fidgets." They flutter restlessly, and Kramer noted they seemed to point repeatedly in the same direction. He wondered if that could be the direction they would fly if they were free. He chose European starlings for his experiments.

By long and careful observation, Kramer charted movements of the caged birds. In the fall the starlings in their nervous fluttering tended to remain in the southwest corner of their cages. This corresponded to the normal direction of starling migration in that part of Germany. In the spring the birds demonstrated a desire to follow the path of northerly migration.

This was only the beginning of Kramer's now-classic experiments. By surrounding the cage with movable mirrors which could change the apparent position of the sun, he determined that the starlings did not know which direction to fly without taking a bearing on the sun. Beyond that, however, once the birds had taken such a bearing, they continued to function for at least six more hours by "memory" of changing angles of the sun even though they could not see it. The starlings behaved as if they had both a compass and a clock as accurate as a chronometer.[10]

Animals also migrate according to certain rhythms geared to feeding and breeding necessities. Some are diurnal (active by day) and others nocturnal (active by night). Behavior during rest and active cycles, once attributed to instinct or acquired habit, has been found to follow precise internal timing.

As an example, Patricia Jackson DeCoursey chose the flying squirrel as her test subject at the University of Wisconsin. She observed that the squirrels naturally leave their den trees at dusk and time their activity almost exactly to sunset as it varies from one time of year to another. In a cage, a squirrel continued to follow his light-dark rhythm just as precisely.

Mrs. DeCoursey moved a number of squirrels into a deep basement where they could be shielded from such external cues as daylight and darkness. In continued darkness, the squirrels drifted away from the 24-hour day-night changes going on outside the building. Different animals adopted days ranging from twenty-two hours, fifty-eight minutes to twenty-four hours, twenty-one minutes, but each individual animal, after it had adjusted to its own particular day, continued to keep precise schedules within this "free running" period. In other words, a particular squirrel might lose or gain time in relation to the outside environment, but each appeared to have an internal clock which regulated its own specific day.[11]

This is only one of many studies in the circadian rhythms of animals. Some hibernate in winter while others aestivate, such as desert creatures which spend their summers in a torpor underground during the hottest and driest part of the year. Although environmental changes could explain a bear's instinct to seek a sleeping cave at the first snowfall, what clocks work while the bear sleeps to tell it when to awaken in the spring? Certain animals apparently have an annual built-in clock which operates in cycles independent of external signals.

Such rhythms were discovered in the ground squirrel by Eric T. Pengelley and Sally J. Asmundson at the University of California in Riverside. They placed squirrels in a windowless room on a schedule of twelve hours of artificial light each day and provided them with unlimited food and water. Regardless of temperature (which was varied in different tests from near freezing to 54° F.) the small animals, without any known environmental trigger, first increased their food consumption and gained weight. Then they went into hibernation, awoke to feeding and activity again for a few months, and again lapsed into hiberation. The period of each complete cycle was a little less than a year. "The animals' behavior fulfilled the criteria for existence of an endogenous (internal) annual clock," the scientists reported.[12]

.. Scientific Experiments in Pinpointing the Biological Clocks

MOST research in biological rhythms has centered upon small laboratory creatures which lend themselves to control and isolation from confusing external influences. Once the existence of internal rhythms is determined, the next step is to attempt to locate its center.

Dr. Janet Harker labored many years at Cambridge University in England with the common cockroach, one of the more unpleasant fellow travelers on the earth, and one of its oldest inhabitants. The reward for her meticulous research in the 1950s and 1960s was the first discovery of the exact location of a biological clock.

The cockroach times its activities precisely. As any housewife knows, it roams the kitchen by night and rests during the day. In a series of delicate surgical operations, Dr. Harker removed one part after another, tracking down, by the process of elimination, the source of the hormone known to be involved in the insect's periodicity. She found it in a cluster of nerve cells called the subesophageal ganglion. This is a secondary brain about the size of a pinhead in the cockroach. Ultimately she tracked the timing mechanism to a group of four nerve cells. Dr. Harker also proved that the time signals of the cockroach are carried through the bloodstream and not by nerve connections, and can be transmitted by body fluids from one insect to another.[13]

Such work is beyond the patience, ability, and imagination of the average person, who may ask the point of it all. Yet it is the history of animal experimentation which has led most clearly to improved health and life for human beings. There is strong evidence that research in biological rhythms will yield beneficial results to humans far beyond present conjecture.

The cockroach is one obscure creature in nature. What of others?

Dr. Robert Y. Moore at the University of Chicago works with rats, which live by regular daily patterns of rest and action even if kept in total light or darkness.

"We have known for some time that these cyclical behavior patterns are triggered by the daily cycles of light and darkness in an animal's environment," he said, "but until recently the exact parts of the brain that triggered these events were not known."

Dr. Moore first blinded several rats by cutting the primary optic nerves running from eye to brain. They still showed normal glandular cycles associated with light and dark. Then he varied the patterns of light and darkness, but the blind rats continued to respond in the same manner as normal animals. With a second group of rats, Dr. Moore cut another set of nerves—the accessory optic tracts running from eye to brain. Now the rats still could see, but severing this special set of ancillary nerves confused their behavior, which finally reversed to the opposite of normal. The rats slept by night and were active during the day.[14]

Scientists still are not entirely satisfied with answers to the question of whether internal clocks are self-operating or must receive some signal from the earthly environment. The answer may be different for each species of plant, insect, or animal.

In a field which contains so many variables, answers emerge only bit by bit as one research scientist or another edges into the unknown, most often by the process of elimination. If it should appear, as it does to many investigators, that unknown and unmeasured external forces are responsible for the performance of biological clocks, such a hypothesis may be proved or disproved only by trial and error. Since to the best of our knowledge all life forms evolved on an earth rotating every twenty-four hours (plus or minus a few microseconds as the ages passed), then logic suggests that the rotation itself and changing magnetic fields might be sensed by a plant or animal, thus forming one of the "windup keys" to the biological clocks.

Karl C. Hamner, professor of botany a. the University of California in Los Angeles, and James C. Finn, Jr., a graduate student, devised an ingenious experiment to test the rotational theory. The two men reasoned that if living organisms could be taken either to the North or South Pole, and there rotated at a speed exactly equal but opposite to the earth's rotation, their internal timing could be tested in a stationary position.

The experiment was conducted at the South Pole using a turntable to cancel the earth's movement. Dr. Hamner's associates worked with hamsters, bean plants, fruit flies, cockroaches, and a fungus. Under all of the controlled conditions, the biological rhythm of the organisms continued as though nothing unusual had happened. As Dr. Hamner commented, the experiments did little to clarify the nature of internal

clocks but they did render "highly improbable" that the
earth's rotation is a triggering force in regulating the clocks.[15]

. . . Studying Circadian Rhythms in Space

TO SOLVE another piece of the puzzle, some unusual little
mice are scheduled to take a ride in outer space. In their
spaceship they will be accompanied by several hundred fruit
flies.

This experiment is being developed by Dr. R. G. Lindberg
of Northrop Corporation's Space Laboratories in California
and Dr. Colin Pittendrigh of Princeton University. The experimental subjects are *Perognathus longimembris*, a tiny desert
mouse which operates by precise circadian rhythms regardless
of light or dark, high or low temperatures, varying humidity,
with or without food, and without water which it does not
require. The mice and fruit flies were selected for the space
experiment because both have been studied extensively on
earth and both live by well-defined 24-hour periodicity.

The Northrop scientists have been busy for the past two
years building special houses to send six of these mice into
orbit around the earth. They will be launched on a 21-day
journey (with a 30-day supply of sunflower seeds) sometime
in 1973 in the National Aeronautics and Space Administration's Skylab program.

Each mouse will be housed individually in a canlike com-
partment which will be kept dark with a steady temperature
of 20° C. The fruit flies will ride in four other containers,
each carrying 180 pupae. As we have seen, the rhythmic hatch-
ing of fruit flies can be modified by a flashing light. This
experiment will be repeated in space to learn if behavior is
different when the flies are removed from earth's gravity and
rotational speed. Also, two different populations of flies will

be hatched at different temperatures to test this factor in relation to internal rhythm and weightlessness.

Each mouse will carry within itself a tiny sensor which will telemeter information back to earth while it is in flight. Developing such instrumentation which could be implanted in the tiny animals without disturbing their physiological functions has been one of the major challenges in the experiment.

The basic idea is to determine if the circadian rhythm of both mice and flies will remain unchanged when they are removed from environmental factors on earth, including gravity. In particular, scientists wonder if the small creatures will remain on a twenty-four-hour daily cycle, or adjust to the ninety-minute cycle of a spacecraft's revolution around the earth.

"If environmental stimuli indeed set the biological clock," Dr. Lindberg said, "then studies of circadian rhythms in a variety of life forms in deep space may be an essential prerequisite to sophisticated manned missions of extended duration." [16]

The lowly fruit fly and a tiny mouse may shed new light on biological rhythms when taken into an environment alien to their ancestral home on earth. They may help us learn how men may live and work during space journeys much longer than the fortnightly trips so far taken in orbit and to the moon.

Plants, animals, and sea creatures all appear to operate by internal biological clocks apparently regulated by shifting tides, light and darkness, and the changing seasons. When removed from natural surroundings, the clocks of many forms of life continue to operate in precise rhythm while others adapt to new rhythms.

Studies with well-organized insects, such as bees, indicate that the biological clocks of life also may be adjusted to adapt to changing environment and location.

Spaceflight experiments with lower life forms will help to prove if the clocks are dependent upon, or independent of, external influences. Such flights beyond the earthly environment may serve as a guide to how well men may be able to adjust to long-term exploration of the solar system and the stars beyond.

CHAPTER 3

The well-hidden physiological rhythms in man. How the human body keeps time despite the conflicts with man-made time.

Man's Master Clock—
Everywhere
and Yet Nowhere...

DEFINING THE INTERNAL CLOCKS of man is more difficult than dealing with plants and animals.

First, our most obvious rhythms, such as waking and sleeping, are accepted without questioning whether we sleep at
the proper time or proper number of hours.

Second, our activities are so confused by habit and artificial
time cycles that natural rhythms may be masked or hidden.
Most people eat three meals a day at specific hours whether
they are hungry or not. Also, most people are locked into a

daily work schedule which disregards individual cycles of efficiency.

Third, an animal or plant can be isolated to observe natural cycles without interference by environmental factors. It is difficult to do this with humans unless they are confined for treatment of mental or physical illness—which, in turn, would confuse natural rhythms.

Fourth, the human receives all his training after birth, functioning with free will and without such instincts as those which guide birds in migration or send a bear into hibernation. It is unlikely, therefore, that any two people have exactly the same natural rhythms.

Despite these difficulties, evidence points to the existence of internal human "senses" which respond to the passage of time without conscious awareness. Such a clock—perhaps tuned to an accumulation of known and unknown environmental influences—may be a central governor of our mental and physical operation. One inkling of the "clock" is found in those people who are able to awaken at an hour preset before they went to sleep.

It appears there are many rhythms and perhaps many clocks under the direction of a master controller which coordinates the ebb and flow of complex functions such as internal secretions, metabolism of food and chemicals, sleeping and waking, fluctuation in mood, and even the division of cells. The biological cycles of a man may range from microseconds to hours, days, seasons, and years, and with little relation to the artificial clocks we use to measure time.

Aside from heartbeat and breathing, the most commonly recognized human rhythm is ovulation and menstruation in women, a cycle geared not to solar days and weeks but to the passage of the moon. An internal clock apparently determines the ripening of an egg, prepares the womb to receive it once fertilized, and directs removal of its protective fluids and substances if the egg is not fertilized. This mystical, emotional, and physical link between the moon and the elemental drive of reproduction has left an indelible mark on fertility beliefs, a mark stronger than rationalization, which tells us that a direct link between human and moon cannot exist. Yet perhaps it does.

All women do not follow the same menstrual cycle, and many fluctuate individually from month to month, but the rise and fall of hormonal tides influence a woman's physiology, emotional responses, mental performance, and even dreams. Around the moment of ovulation some women feel happy and energetic. Before onset of menstruation, many experience special tension which can cause depression, irritability, and even illness. This cyclic depression indeed may play a role in crimes committed by women. One survey of 249 women's prison inmates revealed that 62 percent of their crimes were committed during the week preceding their menstruation.[1]

Men also show monthly changes. Sanctorius, a seventeenth-century physician who weighed himself and others over long periods of time, found that normal men underwent a monthly weight change of one or two pounds.[2]

... The Ups and Downs
in Human Emotions and Efficiency

IN 1887 Wilhelm Fliess, a German physician, published a "formula" for biological rhythms. He had found that children were prone to illness at regular intervals. For twenty years he collected data on thousands of people, charting the ups and downs of their lives—accidents, illnesses, marriage, divorce, pay raises, being fired from a job. The chart lines formed waves rising and falling regularly through the months. From these Fliess identified what he claimed were two basic cycles in human nature, both men and women. The male component (strength, endurance, courage) was keyed to a cycle of twenty-three days. The female cycle (sensitivity, intuition, and love) had a period of twenty-eight days. Both cycles, he claimed, are present in every cell, manifesting themselves in ups and downs of vitality and eventually determining the day of one's death.

Fliess related nasal irritation to neurotic symptoms and sexual abnormalities. He treated illness by inspecting the nose and applying cocaine to what he called "genital cells" in its interior. Most scientists today agree that his treatment methods were false; yet his basic idea of periodic cycles in human function and vitality may have approached the truth.

In 1928 Dr. Alfred Teltscher, an Austrian professor, studied 5000 high school and university students to see if he could find a rhythmic pattern of clear thinking and alertness. His charts corroborated the Fliess twenty-three-day and twenty-eight-day cycles, but he also found a thirty-three-day intellectual cycle of memory, alertness, and reasoning power.

In the 1930s the Pennsylvania Railroad retained Dr. Rex B. Hersey, psychology professor, and Dr. Michael J. Bennett, an endocrinologist, to make a detailed study of skilled mechanics and engineers. Daily records were kept of their conversation, mood, outlook, physical condition, and work efficiency.

No one was looking specifically for biological cycles, but the men's overall ability seemed to fluctuate on an average frequency of thirty-three days. In other studies Dr. Hersey found signs of four- to six-week cycles of emotional change. In some men the shift from an easy-going amiability into a period of tension and irritability came in swings so gradual the men themselves did not notice.

Until recent years only dramatic alterations in mood, such as those suffered by the mentally ill, were considered worth recording. Today, however, it may be possible to measure and tabulate less obvious slow rhythms of emotional transition. Some psychiatric clinics are applying computer analysis to day-to-day staff notes on patients. The data often shows periodic mood shifts that had escaped notice before. The same techniques may show whether healthy people also go through cyclic shifts in mood and physical condition over weeks and months.[3]

A twenty-eight-day rhythm was observed in the drifting bedtime of a physicist during a year's research in Antarctica, where there is no twenty-four-hour change in day and night. Analysts found that he would go to bed later and later each night for twenty-eight days, then revert to his original time of retiring.

Dr. Franz Halberg at the University of Minnesota, world-known for his studies of biological rhythms and mathematical techniques for identifying them, has found evidence of annual cycles in human death. Studying records from the Minnesota Department of Health, he found that fatalities from arteriosclerosis (hardening of the arteries) were highest around January. A peak in suicides occurred around May, and accidental deaths peaked in July and August. While some such cycles may relate to social customs, such as summer vacations, others may correlate with seasonal changes within and help explain certain physical and mental disorders.

Scientists long have been intrigued by "Arctic hysteria," a mysterious psychosis often suffered by Eskimos for a few hours or days during the winter. Dr. Joseph Bohlen and his wife of the University of Wisconsin went to Wainwright, Alaska, and studied ten Eskimos for ten days during each of the seasons. They discovered a wide fluctuation in the secretion of calcium, a chemical which influences the nervous system. The Eskimos secreted eight to ten times as much calcium in winter as they did in summer. That does not prove the cause of Arctic hysteria, but it indicates that understanding of basic cycles may offer clues to many human ailments and disturbances.[4]

. . . Circadian Rhythms in Human Beings

As IN plants and animals, most detectable human cycles are circadian, geared to the rhythm of night and day. Many are explainable, while others seem to bear no relation to known cause and effect. One of the latter was a tabulation of 600,000 normal births showing that most children are born in the early morning with the fewest in the afternoon.[5] No one knows what influence of sun, moon, or stars might relate to such a divergence from the statistical law of averages.

Dr. Mary Lobban of the National Research Institute in Great Britain believes that out-of-phase rhythms within the individual may affect the learning speed of Eskimo children. At Inuvik, in the Canadian Northwest Territories, she found that kidney excretions and alertness among Eskimo children peaked at evening rather than midday, which is normal in middle and southern latitudes. "Because of this evening peak," she said, "it's reasonable to assume that native children here obviously do not achieve their top mental capacity or awareness until long after they have left the classroom for the day."[6] It follows that changing school hours might assist these children to more efficient learning.

We know the vital rhythms of heartbeat and breathing are controlled somewhere in the unconscious nervous system and that abnormal rates are signals of illness. The heart and lungs also are synchronized rhythmically together. A person at rest has a pulse rate of sixty to eighty beats and respiration of fifteen to twenty breaths a minute. The two functions together show a circadian rhythr rising to a peak by day and falling to a low point during sleep. Dr. Gunther Hildebrandt at the University of Marburg, Germany concludes that a heart-beat-to-respiration ratio of four to one is a sign of health. A higher or lower ratio may signal that the body is not functioning properly.[7]

There also are daily periodic variations in other functions. These include sugar content of the blood, endocrine gland and urine secretions, plasma concentrations of certain hormones in the brain and blood, and deposits of carbohydrate in the liver. Some functions reach their peak in the morning, others at night. A healthy person urinates most in the morning with accompanying excretion of certain chemicals. More cells in skin and muscles divide during the night hours than during the day.

Most of these functions occur in timed phase with each other, and when the relationships are disturbed a person may notice it. He may need to go to the bathroom more often during the night, or feel tired at the hours when he normally does his best work. Dr. Beatrice W. Sweeney, research biologist at the University of California in San Diego, compares the intermeshed biological clocks with the automatic timer in a modern kitchen.

"It's as if you had three appliances—a coffee-pot, a broiler and a radio—all plugged into a single timer," she explained. "The coffee pot might be set to go on at 7 A.M., the broiler at noon and the radio at 6 P.M. If something goes wrong with one of the appliances the others will continue on their regular schedule. You can reset the timing mechanism as a whole, but not the interval between events."[8]

The more complex human system must rely upon an orderly sequencing of thousands of events. If, as it appears, a master clock is in charge of the multitude of rhythms in brain and body, no one yet knows where it is or what drives it. As stated by Frank Brown at Northwestern, "it seems to be everywhere, and yet nowhere when we try to localize it."[9]

The clock is not driven by food, work, standard time sense,

or social custom. Internal timing is not precise in newborn babies but adjusts itself after birth regardless of the training program to which the child is subjected. Many circadian rhythms persist no matter what is done to change them. If tissues are removed from the body and kept alive in a nutrient broth, they continue to throb with the same tempo as if they were still part of the body. The rhythm exists even in single cells.[10]

In one experiment, conducted near the North Pole, several men were isolated in a cabin. In addition to the absence of regular light and darkness, their time sense was confused by setting their watches fast or slow. The men worked, ate, and slept in days that sometimes were twenty-two hours and sometimes twenty-eight hours long. Despite this erratic schedule, no change was detected in most of their internal rhythms.[11]

Many years ago Dr. Nathaniel Kleitman, then at the University of Chicago, used himself as a subject to test the circadian rhythm of sleeping and waking. With a young student he spent a month in Mammoth Cave, Kentucky, trying to adjust to a 28-hour day. The student, twenty years younger, adapted to the unusual routine, but Dr. Kleitman reported he was always out of phase, sleepy when he should have been alert and not hungry at mealtime.[12]

Dr. Erwin Bunning of the University of Tübingen, Germany postulates that there must be a rhythmic harmony between cell behavior and the whole organism in health, probably organized around the unit of a day. Frank Brown believes the clock is an "open system" with periodicity persisting through response of the living organism to its geophysical environment. Dr. Harold S. Burr at Yale University suggests that the pattern of a human brain, which regulates and controls it, is a complex magnetic field. It is possible that earthly magnetic fields may somehow influence the rhythmic ebb and flow of human behavior.

The various hypotheses, however, do not tell whether the 24-hour day is the precise rhythm a human would follow were his cycles permitted to "run free." Near Nice, France, two young men lived for two months in a cave without clocks or other time cues. Jacques Chabert, twenty-nine, was exposed to constant light which disturbed his sleep before he grew accustomed to it and settled into a regular 24-hour routine. The second subject was Philippe Engleder, thirty, who controlled his own lighting. After the first thirty days he fell into a 48-hour cycle, working as much as thirty-six hours at a stretch, then sleeping twelve.[13]

A broader experiment was carried out near Munich, Germany, involving 130 volunteers shut in a concrete bunker and deprived of all time cues, regular meals, or sleep patterns. Most settled upon 25- or 26-hour days. Some became completely disoriented. One man adapted to a 50-hour day, working and reading for thirty hours and sleeping for twenty. When he emerged after three weeks he was convinced he had been in the cave for only ten days.[14]

This type of research supports the theory that the ideal work-rest cycle may differ for individuals. Some writers, artists, musicians, and salesmen—not confined to the standard nine-to-five work day of the commercial/industrial world—have found their own best periods of creative efficiency. Some work better in the morning, others late at night.

Some people also find their cycles of dullness and alertness run in series of days, though most have become so adapted to social habit that they are not aware of these rhythms. The writer Thomas Wolfe, for example, reportedly worked standing up in front of his refrigerator, using its top as a desk, in bursts of energy lasting up to forty-eight hours at a stretch. Then he would fall into bed exhausted, leaving scraps of manuscript (and food?) strewn about his room.

Awareness of the passage of time varies widely from one individual to another and often bears little relation to the mechanical clock. Men generally are more conscious of time than women, which may result from different life styles. The average housewife, except for the constraint of preparing three meals a day, lives much by her own time rhythm. Her husband, on the other hand, is regimented to punctuality in business hours and appointments and his habits camouflage his natural rhythms. One might speculate that the woman's greater synchronization with natural time may have a bearing upon the fact that women on the average have longer life-spans than men.

"Many biological rhythms in man, such as body temperature, hormone secretion and blood pressure, vary with the seasons or other cosmic forces," commented René Dubos at Rockefeller University. "Some of man's deepest biological traits are governed by the movement of the earth around the sun, others are connected with movement of the moon around the earth, and still others result from daily rotation of the earth on its axis.

"All of these fluctuations in biological characteristics," he added, "probably derive from the fact that the human species evolved under the influence of cosmic forces that have not changed. These mechanisms became inscribed in the genetic code and persist today even when they are no longer needed under the conditions of modern life." [15]

Although our periodicities may be vestiges left over from evolution, more knowledge may prove the need to pay greater attention to them.

For centuries the biological sciences have been rooted in a natural law which presumed that all functions of the body are held in a steady state of balance, known as *homeostasis,* the French *la fixeté du milieu interieur.* Now we begin to see our interior as a multitude of forces changing in dynamic harmony of intermeshed cycles and rhythms.

"Today we are urged to envisage a milieu constantly changing," stated Dr. J. N. Mills of the University of Manchester,

England, "with a predominantly circadian rhythm but with many components of longer or shorter duration; any observation [of the body] may be suspect without a statement of the time of day at which it was made." [16]

As expressed by Dr. Bertram Brown of the National Institute of Mental Health, while the hours of a day or night flow around us "we perform a complex series of cycles internally, a progression of changes that alters our performance, our senses, the way we metabolize food, the symptoms of our illnesses, our vulnerability to stress or disease, and even some of the subtle displays of vitality and idiosyncrasy we fondly know as personality." [17]

One primary cycle is body temperature, which persistently rises and falls one or two degrees each twenty-four hours.

A person's favorite hours of the day are likely to coincide with the high point, usually afternoon or evening for a person who is active by day and sleeps at night. This cycle may help to explain the difference between people who jump out of bed in the morning, alert and ready for the day's problems, and others who drag around for an hour or two before they begin functioning.

Observations by Lawrence Monroe at the University of Chicago showed that a good sleeper's temperature tended to rise before he awakened in the morning and reached normal at about the time he got up. The poor sleeper's temperature did not rise so early and still was rising long after he was awake. Such results suggest clues to different performance and response between so-called day people and night people.

The day-night temperature cycle persists even when a person is confined to bed, and Dr. Mills points out that the rhythm holds steady whether a person is in a temperate or tropical climate. People whose time phase is shifted need considerable time for their temperature rhythms to adjust. Night workers in a hospital require as long as three weeks to adapt their temperature rhythm, and an equal time to readjust after returning to day work.[18]

At the time of day when our body temperature is reaching its peak, other functions also are changing. The speed of pulse, pulse pressure, pulse-wave velocity, circulating blood volume—all vary by circadian rhythm. The pulse rate seems to be high about the time temperature is highest, and drops during the night, along with blood pressure, in a slow decline similar to the temperature cycle. The excretion of adrenal steroid hormones also is cyclic, remaining roughly in phase with high and low temperatures.[19]

Another group of internal body cycles is connected with the kidneys, which filter poisons and excrete them by urination. As early as 1890 a German investigator named Lahr experimented on himself by staying in bed and taking fluids around the clock. His urine flow remained cyclic, reaching a low during sleep and a high during the day. This pattern of urine flow is persistent regardless of an individual's habits. Even among nurses who were adjusted to night work and slept well by day, urine flow continued lower at night.[20]

The rate of urine excretion may depend upon several other internal clocks. One is the antidiuretic hormone ADH, which comes from the posterior pituitary gland on a pronounced 24-hour cycle of its own. Also excreted in urine is a group of ions, known as electrolytes, including potassium, sodium chloride, phosphate, magnesium, and calcium. Excretion of these substances follows several different circadian rhythms.[21]

It is through the ebb and flow of these substances that investigators obtain vital clues to the periodic waves of activity going on in the body. Creatinine, a by-product of muscular activity, varies by circadian rhythm as do many hormones, including those which influence sexual function.

The same is true of the blood, which contains many varieties of cells. These exhibit 24-hour fluctuations of quantity often differing from each other. Measuring the proportions of different blood cells at different times of day can indicate what is happening in the cortex of the adrenal glands.

Other properties of the blood, including the ability to clot and levels of sugar, change circadianly. One symptom is low energy and irritability suffered by some people whose cycle of blood sugar hits a low point before dinner in the evening. This cycle seems related to the liver, which stores glycogen, a basic energy substance, and regulates the blood sugar (glucose) level. It may be rhythms in the liver, rather than an empty stomach, which control our periodic hunger for food. If the liver is not functioning in harmony with other cycles, it may allow ammonia to pass through the bloodstream to the brain. This can cause temporary psychosis.[22]

There also is circadian periodicity in the production and breakdown of the body's most fundamental energy unit, adenosine triphosphate (ATP). Studies by Dr. Pittendrigh at Princeton revealed that ATP is transformed to release energy 25 percent faster during time of activity than during rest.

Importance of this cycle was demonstrated at the Lafayette Clinic in Detroit, where a group of people were deprived of sleep. After being awake for 100 hours, the volunteers began to suffer hallucinations and psychosis. They saw fire bursting from the walls and suspected their friends of conspiring to kill them. Their muscle performance and coordination also declined. Blood samples indicated that sources of the basic energy unit were beginning to run down. Once the subjects returned to their normal rhythm of sleep, energy production and behavior also returned to normal.[23]

This relates to one of the most essential rhythms in man's internal day—what we eat, how much, and when. Most people bolt a light breakfast, eat a variable lunch, and then enjoy a heavy dinner in the evening when relieved of the day's tensions. This schedule may be opposite to the body's natural rhythms, because protein-rich foods produce greater nutritional value at one time than at another. The reason is that the enzyme which converts amino acids into body fuel functions on a 24-hour cycle.

Dr. Richard J. Wurtman of MIT and Dr. Julius Axelrod of the National Institutes of Health found in animal tests that protein utilization fluctuated as much as 400 percent from one part of the day to another. Best use was made of protein consumed in the morning and there were indications that protein consumed in the evening may be partially wasted. The best cycle of protein conversion may result from eating the largest meal at breakfast, a lighter lunch and the lightest of all at dinner.[24]

Internal clocks also govern sensory keenness, which varies during the day, possibly in synchronization with blood levels of hormones. This cyclic change was found originally in patients suffering from insufficient output of adrenal hormones either because of disease or problems in the pituitary gland. In general, it appears that the person with low output of adrenal cortical steroid hormone is highly sensitive to all stimuli. Sensory perception in normal people also fluctuates with the rhythm of adrenal hormones. A person on standard sleep schedule will reach a high point of sensory sharpness about 3 A.M., when steroid levels are lowest (and the neighbor's dogs are barking). After that there is a drop in sensitivity as steroid levels rise. The senses then become gradually more acute again, reaching another high point around 5 to 7 P.M. It is no accident that the evening meal may smell and taste better than any other meal of the day.

The pace of cell division in organs and body tissues also shifts according to changing hours of day and night in a pattern of rhythms coordinated with other cycles in glands and vital life centers. Dr. Halberg at Minnesota also found 24-hour periodicity in the activity of the fundamental nucleic acids, DNA and RNA, which transmit patterns of heredity and govern protein production and cell division. DNA and RNA do not reach their peaks at the same time and, in the liver, seem related to rhythmic increase and decrease in levels of fats and sugars.[25]

There is considerable difference in individual biological time clocks. Most people are not aware of their natural rhythms unless placed in isolation and allowed to run free. In others, however, the sense of internal time is so strong that they live by their own schedules rather than the artificial clock time of society. One such is D. R. Erskine of Philadelphia, who followed his own internal time for fifteen years. He found his greatest sense of well-being by retiring sixty-five to eighty-five minutes later each night and rising that much later in the morning. This threw him out of phase with society, forcing him to consult a calendar to schedule himself to a social event, but he reported his greater enjoyment of living made his own schedule worthwhile.[26]

.. The Trigger of Light

RUNNING a steady current through all studies of biological time clocks is the apparent fact that they are set—or may be reset—either by inherited predisposition or by reaction to contemporary forces such as light, temperature, or geomagnetic forces from the earth, sun, and moon. Such time-givers or triggers are known by the German word *zeitgeber*. According to Dr. Mills at least forty different circadian variations exist in the environment. Thus in any given situation multiple zeitgebers or time-setters may be at work, or a single one may be dominant.

One of the most important is light, particularly light of day. This varies widely with the seasons and also in day length from the tropics to the polar regions.[27]

We have seen how changing hours of daylight affect different plants, as in the flower clocks of Europe. Changing light of the seasons also affects plant and insect rhythms in predictable ways learned over a thirty-year span by Dr. Bunning of Germany. He demonstrated that solar rhythm does not persist exactly when plants are kept in total darkness. In zeitgeber experiments he found he could give plants ten hours of light and ten of darkness, after which they would continue a twenty-hour cycle of leaf movement. He discovered also that plants could adapt to many different light-dark cycles, but when left in total darkness, they reverted to a circadian rhythm of not quite twenty-four hours. When he raised seedlings in darkness, they exhibited no rhythmicity at all but a single momentary exposure of light was enough to start their circadian clock functioning again. Apparently plants inherit some kind of time-map of their environment which gives them flexibility but also prepares them for seasonal changes.

Dr. Bunning points out that living things not only *indicate* the time of day but also make use of the clock to *predict* future events. Such is a flower which begins opening before dawn, as though anticipating the sunrise, and then is secreting nectar with a "time sense" synchronized to the arrival of bees which are necessary for pollination. The bees train their own internal clocks to be at the right place at the right time to reap their honey harvest.[28]

Birds and mammals also exhibit light-sensitive cycles which appear to dominate other signals. Without internal timing linked to a common zeitgeber, many birds and fish would not develop simultaneously in a manner permitting them to reproduce or to gather at spawning grounds in large numbers at the right time.

Dr. Jurgen Aschoff of the University of Munich, Germany found that if finches (which are active by day) are kept in constant dim light with only fifteen minutes of bright light a day, their activity cycle could be influenced by the timing of the bright light.[29] "Aschoff's Rule" states that light accelerates the biological clocks of creatures which normally are active by day but delays the cycles of nocturnal creatures. Light apparently prepares all living things for the coming season. As once stated by Dr. Joseph Meites of Michigan State University:

"In the spring a young squirrel's fancy turns because the days are getting longer, and exposure to longer light periods sets off a chain-reaction involving the brain and pituitary glands, resulting in releases of hormones that affect sex hormone levels and in turn cause the sex glands to enlarge and produce their sex hormones." [30]

In 1920 researchers captured migrating finches in the Canadian province of Saskatchewan as they were heading south. The birds were held in zero cold in outdoor aviaries, but at sunset of the shortening days they were exposed to light from electric bulbs for a few minutes longer each day. By mid-December the birds were singing mating calls and the testes of the males had developed. When they were released in mid-winter in Canada, they headed north instead of south.

In animals light stimulates secretions by the hypothalamus and pituitary gland, which in turn increases estrogen secretion by female ovaries and the male's production of sperm. Changing wavelength of light apparently is important also, and even blind animals respond to light if certain nerves remain to transmit the radiation to a central switching point in the brain. This switching point, or central clock coordinator, has been tentatively traced to the pineal gland.

This gland, shaped like a tiny pinecone, is situated deep between the two hemispheres of the brain. Considered historically as a vestigial third eye, it is sensitive to light. The pineal and its extracts apparently influence the thyroid glands, alter the size of the adrenals, modify brain-wave tracings, induce sleep, and influence rhythmic development of the sexual glands. Most scientists regard the pineal gland as a coupling device which may regulate the phase relations among many rhythms, using light information to govern changes.

In man there are indications that light may trigger an increase of adrenal hormones. In this way blood hormone levels would rise in the morning in *anticipation* of wakening and activity rather than rising in response to these. The pineal gland's signals of light change may be one mechanism in human seasonal changes such as "spring fever," the periodic increase and decrease in suicides, and variations in some illnesses such as ulcers and psychoses. These signals may have other more subtle effects as well.

In 1964 a young French woman, Josy Laures, lived in an underground cave for three months to test her free-running cycle. According to Dr. Reinberg of Paris, while she lived in isolation her menstrual period shortened from twenty-nine to twenty-five days. Dr. Reinberg speculated this change may have been triggered by the fact that her only light underground was a dim miner's lamp.[31]

In a study of 600 girls from northern Germany, Dr. Reinberg found most of them entered their first menstrual cycle in the short, dark days of winter. Such studies suggest that dim light, or perhaps lack of light such as in the blind, may stimulate the reproductive hormones in women. Relationship between the pineal gland and sexual maturity also is indicated by the fact that children with pineal tumors reach puberty grotesquely fast, some as early as kindergarten age.

It was Dr. Reinberg and a group of physicians in New York who in 1967 discovered a daily testosterone (male hormone) rhythm in men. Inasmuch as sex hormones frequently are used to treat cancer of the uterus, breast, and prostate, knowledge of this natural rhythm may make more precise treatment possible. Other studies have shown that exposure to light at certain times can either delay or speed up ovulation.

In 1964 Dr. Edmond Dewan, a physicist at Cambridge Research Laboratories in Bedford, Massachusetts, studied a thirty-six-year-old woman whose menstrual cycle had varied between twenty-three and forty-eight days for sixteen years. During the experiment (though she was not aware it *was* an experiment) she slept for two nights in the indirect light from a 100-watt bulb placed on the floor at the foot of her bed. The light was on only during the nights between the fourteenth and fifteenth days and the sixteenth and seventeenth days of her current menstrual period, flanking the time when ovulation could be expected to occur if her period were more regular. During her first three monthly cycles under this treatment the woman's menstrual rhythm settled regularly on twenty-nine days.[32]

Although we do not yet know how light may reach the brain when the eyes are closed, other tests have fortified the theory that a dim light shown at appropriate times can alter the period of ovulation. Changing moonlight may even be sufficient to trigger the delicate internal clocks.

It is here that we come full circle to the scientific gray areas suggesting it is no accident that the menstrual cycle coincides with the moon's revolution around the earth. We may yet find that the moon exerts influences upon us which are not yet known or measured. There may prove to be critical periods in the cycles of our brain, nerves, and glandular system during which light is as important as it is to the flowering of plants.

> *Although free will, intellect, and social habits tend to disguise the fact, human beings, like lower life forms, function by complex interrelated biological clocks which dictate the true rhythms of life.*
>
> *These rhythms range from long high*

and low cycles of physical and emotional vitality to a daily ebb and flow of internal processes ranging from glandular secretions to the metabolism of food and division of cells. Scientists have reached a number of tentative conclusions which may serve as future guides to greater personal well-being and satisfaction from living. These include:

1. If a person were allowed to "run free," his greatest efficiency might be achieved by adapting to a day shorter or longer than twenty-four hours and perhaps out of phase with standard social adaptation to day and night.

2. Natural daily fluctuations in temperature may signal an individual's natural rhythm.

3. Knowledge of individual internal clocks may help us derive greater value from food and take advantage of high points in sensual acuity.

4. Operation of our biological clocks may be triggered by heredity or by environmental factors, the strongest of which appears to be the changing pattern of natural light.

5. In human beings, intensity of light may play a role in development of sexual functions and influence regularity of ovulation and menstruation in women.

Refining knowledge of this function of light might enable women to determine the exact moment of ovulation and provide a valuable new tool for natural control of fertility and population growth.

CHAPTER 4

*Letting the body tell
when it's ready for sleep,
the brain's time for exercise
and homework. Effects
of sleep deprivation.*

Sleep: Physician, Repairer, Nourisher of the Brain

Sore LABOR'S BATH, balm of hurt minds . . . chief nourisher at life's feast."

That is sleep in the words of Shakespeare's Macbeth, who described the restorative powers of this commonplace yet mysterious process more succinctly than anyone before or since.

The average person spends one-third of his life in sleep. It comes easily and blissfully to the young, bursting with physical and mental growth; but restlessly, with greater difficulty and shorter hours, to the aged in whom the clocks of life are beginning to run down.

A child resists bedtime because he is sad at the loss of a happy day and cannot visualize tomorrow. A young man or woman may regard sleep as wasted time interrupting the urgent need to experience the world. For an adult, sleep may be the clean healer of sore mind and muscle at the end of a day's productive work, or perhaps an interlude promising tomorrow will be better than today. The insomniac woos sleep fearfully and desperately as slow hours of night move by. A hurt and confused being seeks sleep as oblivion, sometimes to the point of praying tomorrow's dawn will never come. Such a prayer may be tragically abetted by an overdose of sleeping pills to end the rhythms of life forever.

Sleep is such an all-pervasive part of the twenty-four-hour cycle that we accept it without question, but it is not the simple process most of us consider it to be. The restoration of body and mind during this essential third of our lives occurs at a multitude of rhythmic levels. Our minds and bodies continue active during sleep, including periodic dreams which are more than nightly entertainment (or horrors) and apparently play an essential role in mental health.

Without sleep a man performs poorly at his work, cannot sustain concentration, becomes irritable, and eventually begins to suffer from hallucinations which result in deranged behavior. If a person abruptly and spontaneously exchanges night for day, waking by night and sleeping in the daytime, it can be a sign of serious illness such as encephalitis, which damages important centers in the brain. A man not only needs rest, but he apparently operates best if he sleeps part of every twenty-four hours.

Importance of this circadian rhythm in mammals was tested by Dr. Curt P. Richter at Johns Hopkins University. He tried without success to alter the twenty-four-hour cycle of rat activity by assaulting animals with shock and drugs, exposing them to prolonged danger, freezing them, stopping the heart, and removing certain portions of the brain. Even after they were blinded, rats persisted in a cycle which differed little from twenty-four hours in full period. The circadian rhythm of sleep and action seemed to be a permanently fixed characteristic.[1]

.. The Cycles of Sleep

SIGMUND FREUD, pioneer psychoanalyst, sensed intuitively that sleep and dreaming are important to mental health. Modern sleep research, however, was begun in the 1920s by Dr. Kleitman of the University of Chicago, who, we have seen, tested some of his own theories in a cave. His pioneering techniques, used much in the same fashion today, involved rigging experimental subjects with electrodes to measure brain waves, eye motion, and muscle tension while they slept. Tiny electrical impulses are amplified and recorded on polygraph paper so that investigators may study the record and compare it with others.

An early discovery was that whether or not we sleep well, we do not sleep in a uniform manner through the night but in regularly recurring cycles of 90 to 120 minutes each, with marked stages in each cycle. Although the governing internal clock has not been located, these cycles have been confirmed by thousands of observations during the past twenty years.

In a typical test, the volunteer sleeps in a comfortable bed with electrodes connected around his head. As he becomes drowsy, the normal waking "alpha" brain waves of nine to twelve cycles a second are replaced by a rapidly changing irregular wave. Breathing is regular, the heartbeat slows, and the body temperature starts to drop. The eyeballs roll slowly. As he drops off to sleep, the person experiences random visual and auditory impressions and thought fragments. This is sleep Stage 1, which lasts only a few minutes.

In the sounder sleep of Stage 2, the eyeballs stop wandering and brain waves shift to twelve to fourteen cycles a second. The subject has started the period which makes up about 50 percent of all nocturnal sleep. Then the electroencephalogram shows the brain waves growing larger and slower, and the sleeper is in Stage 3, which phases gradually into Stage 4, the deepest sleep of all.

By the end of some ninety minutes, the sleeper has returned almost to the surface, as though he were about to waken. Sometimes he does. More often his chin-muscle tension drops to nothing. Brain waves appear almost the same as those of a person awake. Then the two pens attached to electrodes near the eyes reveal that both eyes are darting rapidly, in the same direction, as though the sleeper were watching cars whiz past him.

Because of the *rapid eye movement*, this is known as REM sleep. It lasts for only a few minutes. It was first discovered in 1952 when a Chicago experimenter thought his recording machine had broken down but then discovered the motion of the eyes. When awakened from REM sleep, virtually all test subjects report they were in the midst of vivid dreams.

Other things also happen during this period. Heart rate, blood pressure, and breathing become irregular. Body temperature rises. The kidneys make less but more concentrated urine. Nightmares occur. Males of all ages have penile erections, a sign that a brain center has been activated. During REM sleep, duodenal ulcer patients awaken with stomach pain because the stomach secretes more gastric fluid. Sometimes the jaw muscles contract to cause tooth-grinding. (One might expect other nocturnal disturbances to occur at this time, but it is in deep sleep that researchers have observed such irregularities as bedwetting, sleepwalking, night terrors, snoring and —in children—head-banging.)

After remaining in the REM or dream state for from ten to twenty minutes, the sleeper will slip back into Stages 2, then 3 and 4, and back to 2. The second episode of dream sleep starts about 90 to 120 minutes from the first. Throughout the night a healthy person will sleep through four to six such cycles.[2]

.. The Body's Need for Sleep

THE amount of sleep we need each day varies greatly from youth to old age. New-born babies sleep sixteen to twenty hours a day, decreasing to fifteen by the time they are sixteen weeks old. Through a lifetime, the human being spends 33 percent of his total time asleep. (By comparison the cow sleeps 3 percent, the horse 29 percent, and the gorilla 70 percent of each twenty-four hours.) As a person grows older, the amount of sleep grows less. Men and women from sixty-four to eighty-seven years old average only five and one-half hours a night.

Since there are people of all ages who seem to need less sleep than others the question has been raised if it is essential that we sleep, or if it is a habit which could be altered by training. To solve part of the question, researchers have tested reactions of people deprived of deep, or slow wave, sleep.

Dr. W. B. Webb of the University of Florida deprived volunteers of Stage 4 slumber for seven nights. Although they were able to perform various tasks, the subjects became physically uncomfortable, withdrawn, less aggressive, and concerned with vague physical ailments. When allowed to resume their normal pattern, the volunteers spent more time in deep sleep, indicating the body has a periodic need for it. Importance of regular deep sleep is suggested by the fact that alcoholics, schizophrenics, and patients with chronic brain damage get little of it.

Long-term deprivation of *all* sleep causes more or less

serious consequences depending upon the age and physical condition of the persons involved. During World War II Dr. D. B. Tyler at the California Institute of Technology studied the effects of total sleeplessness on 350 men aged seventeen to thirty-five. The test ran for 122 hours—almost exactly five days.

Many of the men quit before the period ended because they were fearful of bad work performance or of "losing their minds." Of those remaining, four developed hallucinations—seeing or hearing things which did not exist. One talked to unseen persons. Another insisted he heard dogs barking when there were no dogs. Three of the four experienced abnormal false beliefs. A quiet young marine marched away from his group, believing that he was on a secret mission for the President of the United States to seek out spies and traitors.

Mental disturbance from sleeplessness varies from person to person and according to time and physical condition. One man, forced to drive an automobile for sixteen hours to his father's funeral after a full day's work was certain he saw trees converging upon the highway in the middle of the night in treeless eastern Colorado. In January 1959 Peter Tripp set up a booth in Times Square in New York and broadcast for the March of Dimes. He went without sleep for 200 hours. As time passed, he began to "see things." A man's tweed suit seemed to be made of furry worms A desk drawer seemed to spout flames. In 1968, Dr. M. Vojtechovsky, of Charles University in Prague, Czechoslovakia, kept alcoholics awake for 127 hours. Of twenty men in the test, two became temporarily insane.[3]

Keeping a person awake and otherwise disrupting his natural biological rhythms is a basic tactic in forcing him to change attitudes and beliefs as shown by the brainwashing of American prisoners during the Korean War. This technique uses three phases. First, the victim is kept awake in isolation and becomes disoriented and disillusioned. Second, he is questioned at length and at all hours of day and night to deepen his confusion. Third, after he has been deprived of regular meals and sleep, of companionship and reading materials, a man becomes depressed and loses his sense of personal identity. It is at this point that he is ready to accept almost any human contact, regularity, and set of political or moral convictions.[4] By such a simple disruption of normal rhythms, many American prisoners of war were induced to sign confessions opposite to their moral and ethical beliefs.

Far more people, however, are subjected to the many disorders which alter rhythms of slumber in a so-called normal lifetime.

. . . Sleep Disturbances

SLEEP disturbance may be related to illness, such as mental disorder, asthma, heart disease, drug or alcohol addiction. Healthy children may develop teeth-grinding, bed-wetting, body-rocking and head-banging, sleepwalking, and nightmares. Doctors at Cedars-Sinai Medical Center in Los Angeles studied sleepwalking and established that 15 percent of all children between the ages of five and twelve have experienced it. Another study of 500 children between four and fourteen showed that 22 percent had suffered bed-wetting.

Night terrors also are common among children. In such an episode, a child may suddenly sit up in bed, sobbing or screaming. His eyes are open and he seems terrified. He may sit in this manner from ten to thirty minutes and resist all efforts to waken him. Eventually he stops crying and goes back to sleep, and seldom remembers the episode in the morning. Although few treatments help these childhood disturbances, they usually are outgrown.

A more serious problem in the circadian rhythm of sleep is narcolepsy, in which a person suffers irresistible "attacks" of sleep during the daytime. These abnormal naps usually last for about fifteen minutes. Narcolepsy often appears during adolescence or young adulthood and is twice as common among men as in women. Sufferers may fall asleep at the table in the middle of dinner or even while walking along the street, bumping into people as if they were drunk or under the influence of drugs. The cause of narcolepsy is unknown but several drugs are used to relieve the symptoms.

Disruption of standard sleep patterns often accompanies sickness and is one of the most sensitive indicators of the severity of illness. Persons suffering from mental depression sleep less, take longer to fall asleep, awaken frequently, and have more sleep stage changes. Mentally depressed patients enjoy less, or a complete lack of, slow wave or sound sleep. Chronic schizophrenics have a regular amount of dream sleep but only half the normal deep sleep. Mentally retarded people, on the other hand, sleep an average amount of time but spend less time dreaming.

Epileptic seizures involving the brain often occur during sleep, but the attack is not accompanied by the associated spasm or "fit" of wakeful seizures. According to one doctor, it is as if the body were disconnected during sleep from an electrical storm going on in the brain.

Angina pectoris, sharp pains in the chest when the blood supply to the heart is suddenly shut off, often attacks during the dream stage of sleep. In such cases, usually five or ten minutes after dreaming begins, the heart rate increases and breathing becomes irregular. The patient wakes with severe chest pain. As mentioned before, the rhythmic periods of dreaming also coincide with increase of stomach acids in ulcer patients. A test of one ulcer sufferer showed his gastric acid secretion increased five times during REM sleep, whereas no increase at all was detected in normal subjects.

Addiction to drugs or alcohol causes serious disturbance of essential rhythms. Dr. A. Kales of UCLA reported the case of a twenty-three-year-old man who had been addicted to pentobarbitol (a sedative) since age thirteen. Before withdrawal from the drug he enjoyed little sound sleep and reported dreaming only once or twice a year. When the drug was removed his dreaming increased sharply as though his internal clock were making up for lost time. His slow wave sleep, however, did not return to normal until two months after withdrawal from the drug.

British doctors demonstrated with volunteer medical students that drugs affect sleep in two ways, one immediate and the other long-lasting. The subjects took sleeping pills every night for two weeks. At first they reported a decrease in dreams and suffered hangovers each morning. As they became accustomed to the drug, their hangovers lessened and sleep became more standard. When the drug was withdrawn the two students took longer to get to sleep, spent an unusual amount of time dreaming, suffered nightmares, and felt shaky all day. It required five weeks for the two to recover completely from the sleeping pill "habit" of only two weeks' duration.

Similar experience is encountered by those addicted to stimulant drugs, such as the amphetamines, which are favorites among young drug cultists. An immediate effect of taking a standard dose of amphetamine is to reduce the REM cycle. When the drug is withdrawn, total sleep time increases but there is a 50 percent jump in time spent in vivid and unpleasant dreams. This is one of the marks of drug withdrawal. It requires from three to eight weeks for a sleep pattern to return to normal after drugs are discontinued.

Chronic use of alcohol is even more damaging to biological rhythms. Testing chronic alcoholics at the Navy Medical Neuropsychiatric Research Unit at San Diego, Dr. L. C. Johnson found that excessive use of alcohol for ten to twenty years not only resulted in body damage but also totally disrupted the sleep patterns of chronic drinkers. Alcoholics ˏdo not enjoy a stretch of seven to eight hours of sleep a night. Instead, their sleep is broken up into short periods with many brief dream episodes. They seem unable to remain in the REM stage of the cycle and, instead of drifting into another stage, often awaken during the night.

When alcohol is withdrawn, the sufferers immediately enjoy more dream sleep, one of many indications that this phase of sleep in rhythmic patterns is a vital human need. "Enjoy" is perhaps the wrong word to use in connection with the improved REM sleep in withdrawn alcoholics because the dreams are unpleasant and part of the overall discomfort of "kicking" the habit. Dr. Johnson also found that slow wave sleep—almost totally absent among alcoholics—was slow to return after withdrawal. Excessive drinking during a long period of time almost permanently damages the internal clocks which govern the rhythm of slow wave sleep.[5]

Sleep research thus is casting new light upon the harm caused by drug and alcohol addiction in adults as well as the young.

. . . Insomnia

THE most widespread sleep affliction of all is insomnia, the inability of healthy people to obtain their needed quota of sleep in regular circadian cycles. Seriousness of this problem is shown by the fact that Americans spend more than $100 million a year for sleep-inducing prescription drugs, not counting millions more spent for over-the-counter preparations.

The primary problem most investigators have found is that most people who claim to suffer chronic sleeplessness are not insomniac at all. For example, Dr. W. W. K. Zung of Duke University studied the brain wave sleep patterns of a woman who claimed a history of twenty-five years of insomnia. In the laboratory, the woman had no difficulty falling asleep and slept an average of eight hours a night, although her sleep stage rhythms were unusual. During the four nights she remained under observation, the woman claimed she had not slept a wink.

Dr. William C. Dement, one of the leading sleep researchers in the United States, believes that most so-called insomnia is induced by an emergency, special mental tension, noise, or other outside influence, and then perhaps becomes ingrained as a habit. A neurophysiologist who now heads the Sleep Disorders Clinic at Stanford University Medical Center, Dr. Dement has studied most aspects of sleep and is attempting to establish a precise definition of insomnia.

"The term is easily understood in a general sort of way," he said, "but when we attempt to bring some precision into our understanding of the condition, we find it peculiarly elusive.

"We must keep in mind the possibility," Dr. Dement added, "that some patients might be 'cured' if confronted with their own sleep recordings which show they are not having problems with sleep. Generally such insomniacs will sleep longer in the laboratory than they claim to sleep at home."

Dr. Dement pointed out that many mothers suffer temporary insomnia after giving birth to a baby. He thinks this is nature's way of making sure the mother hears her child if it cries at night. Such cases often are treated by physicians who prescribe barbiturates to help the mothers sleep for recuperation. Soon the women develop drug tolerance and the dose is increased. Dement has treated a number of such cases by withdrawing them gradually from drugs, after which the women sleep normally. Some drugs have been found to suppress the dream cycles.

"This is why insomniacs who are given drugs for prolonged periods should be given one of those drugs which produce normal sleep," Dr. Dement said, "rather than medications which suppress or eliminate the REM phase."

There are many true insomniacs, but Dr. Dement believes others could cure their own sleep problems by paying careful attention to their natural internal clocks.

"One of the first things we try to ascertain in the clinic is whether or not a patient is going to bed too early," he said. "This can be done by temperature measurement with a special thermometer. It is easier to go to sleep when the body temperature is on the downward phase of the daily cycle. This is a sign that the body is at its most inactive state and ready for sleep. It is likely that some insomniacs may improve their sleeping habits by delaying their bedtime." [6]

Dr. Paul Naitoh, of the Navy Medical Neuropsychiatric Research Unit at San Diego, feels that the poor sleep suffered by millions of people may be at least partially due to the way we live in our technological world.

"Perhaps the ordinary insomniac—healthy people who just can't sleep—may be the result of our modern society," he said. "As we live and work today, physical labor and exhaustion are rare for many of us and our nutrition is good. The need to sleep to recuperate from the fatigue of the day's work may be much less than it was a hundred years ago.

"Insomnia could stem from our failure to recognize that need for sleep has become less in the 20th century and from sheer habit we are trying to sleep in the absence of any dire need to sleep. Perhaps all of us are not sleeping as well as we think." [7]

. . . What Dream Sleep Does for the Brain

MODERN research has shown that sleep—of whatever number of hours is needed by each individual—is an essential factor ordered by our internal clocks to maintain the rhythms of life. This appears especially true of the mysterious process of dreaming, which has captured attention of sleep investigators of all ages. Freud linked dreams with the sexual drive. Today we are learning that the rapid eye movement, or dream stage, of sleep may be an integral part of the healthy functioning brain. The other stages, lumped together as slow wave, seem necessary for physical rebuilding. Growth hormones reach their peak in young people during slow wave sleep and may continue in adults as part of the physical repair process.

What is the function of the dream stage? We have seen that it occurs for a few minutes in each 90- to 120-minute sleep cycle of all normal human beings so far tested.[8]

As early as 1959 Dr. Dement, then at Mt. Sinai Hospital in New York, sought some of the answers to the function of dreaming by depriving human subjects of REM sleep. Each time a sleeper slipped into the REM stage he was awakened. Then he would skip dreaming and go into another stage, the body apparently giving up temporarily the idea of shifting into REM sleep.

On the first night the sleeper needed to be awakened only a few times, the body unconsciously cooperating in skipping that phase. But after fifteen nights of dream deprivation, his body fought so desperately to get in some dreaming that he had to be wakened on an average of nineteen times an hour. After each arousal his body tended to go quickly into another REM period, indicating a strong need for this type of slumber.

The failure to dream causes a number of changes, including increased appetite, tense aggressiveness, and excitability. Some researchers report that several nights without dream sleep results in the subjects becoming poorly organized and less effective in working with other people.

By taking frequent blood samples scientists have discovered some of the things which happen in the body during REM sleep. Chemical changes include an increase of internal oxy-

Sleep Chart
The Insomnia Beater

To find the hour at which you should go to bed in order to enjoy the soundest possible sleep, first determine your daily average body temperature as follows: take your temperature at 8 A.M., noon, and 4 P.M. daily for a week. Then perform the mathematical operation shown below. It is most likely that your body's daily average temperature is below the 98.6° F. used medically to indicate whether you are running a fever. It may be one or more degrees lower. In any case, determine this average on days when you are healthy. Once you have found your daily average body temperature, determine the hour in the evening at which your body temperature usually begins to fall below that level by filling in the chart on the opposite page. That is the hour at which you should be able to fall asleep most easily—and sleep most restfully through the night.

MY DAILY AVERAGE TEMPERATURE

DAY	1	2	3	4	5	6	7
8 A.M.							
Noon							
4 P.M.							

Total of all temperature figures = _____

Divide total by 21 _____ ÷ 21 = _____

= DAILY AVERAGE
TEMPERATURE

Evening Body Temperatures
(28-DAY CHARTING)

DAY	1	2	3	4	5	6	7	8	9	10	11	12	13	14
8 P.M.														
9 P.M.														
10 P.M.														
11 P.M.														
MIDN.														
1 A.M.														
2 A.M.														

DAY	15	16	17	18	19	20	21	22	23	24	25	26	27	28
8 P.M.														
9 P.M.														
10 P.M.														
11 P.M.														
MIDN.														
1 A.M.														
2 A.M.														

Time at which temperature usually starts
to fall below daily average: _____

= BEST HOUR TO RETIRE

To find your best hour for retiring, make a chart like the one above. Over a period of 28 days take your body temperature every day at the times indicated on the chart and enter those temperatures in the appropriate boxes on the chart. When the chart has been filled in, note the hour at which your body temperature usually begins to fall below its daily average, as calculated by the method shown on the opposite page. This is the time at which you should retire in order to enjoy the soundest sleep.

gen and a decrease of oxygen in expired air. There is lower urine volume and increase in some hormone secretion. The level of free fatty acids in the blood plasma is usually higher in dream periods. Many drugs reduce the amount of dream sleep. Others, such as aspirin and caffeine, do not seem to alter it. A few, including reserpine and LSD, seem to lengthen the dream time. In general, the dream state of sleep apparently is an easily disrupted but essential rhythm of life.

Female rabbits enter the dream state readily and remain in it longer just after copulation. The same does not apply to male rabbits. In humans, dream time is less in women during the first two weeks of the menstrual cycle than toward the end of it. Increased levels of sex hormones, progesterone and perhaps estrogen, are believed responsible. Administration of the male hormone testosterone to elderly men raises their customary low levels of dream time. As Freud perhaps foresaw instinctively, dreams seem to be at least partially related to the release of sex hormones.[9]

A number of theories have been proposed to define the function of dreaming in the 90-minute cycles of sleep. Dr. H. P. Roffwarg of Montefiore Hospital in New York believes REM sleep may help mature the brain in the young. During this period, he says, we receive much visual and auditory stimulation, which is exactly what the brain needs to mature.

Another theory states that dreams provide a "shot in the arm" for the brain when it is deprived of all sensory impressions and reaches an undesirable low in excitation during the other stages of sleep. This period of invigorating stimulation supposedly brings the brain up to a level that conditions it for full alertness upon awakening in the morning.

Dr. E. M. Dewan of the Air Force Cambridge Research Laboratories in Massachusetts sees dreaming as the time when the brain, like a digital computer, organizes or programs itself to cope with the demands of the following day. This view is supported by the fact that the human fetus spends 50 to

80 percent of its sleep in the REM stage. Other scientists consider dreaming to be a time when emotionally meaningful experiences are combined with old memories to complete an up-to-date file of memories in the brain.

Dr. F. Snyder of the National Institute of Mental Health suggests dream sleep is a vestige left over from a vital process to insure survival of the species. When mammals first evolved, some 200 million years ago, they were surrounded by hordes of gigantic reptiles. It was not an easy life for the mammals, which lived under constant threat of being caught and eaten. The longer the small animals hid in caves or crevasses and slept to conserve their energy, the better their chances of survival. Sleep, however, makes an animal helpless to meet dangers or emergencies unless a sentinel is posted. Dr. Snyder speculates that early evolution in mammals included a built-in clock that awakened them periodically— about every hour and a half—for a quick check of the environment, enabling them to detect danger without completely disrupting sleep. The dream period does just that since about 85 percent of all REM sleep is followed by a brief awakening. Supporting this hypothesis is the fact that during dream sleep, the body is roused sufficiently that, if a brief awakening did reveal danger, the animal could immediately spring into action to escape it.[10]

Dr. Ian Oswald, senior lecturer in psychiatry at the University of Edinburgh, Scotland, has placed the function of our two major kinds of sleep into logical perspective based upon long years of experimentation. He points out that in order for the body to restore itself after a day's work, new cells must be produced to replace those worn out or destroyed. Also, although we do not keep forming new brain cells, their structural components and working machinery do not last indefinitely and there is always turnover of these components. Old items are replaced by newly synthesized ones.

"It now seems likely," Dr. Oswald said, "that sleep somehow assists in these two types of synthetic process for growth and cell renewal and two kinds of sleep help in different ways."

He pointed out that growing children have more slow wave sleep than adults and older people. The same is true of athletes after a day of hard exercise and in people with an excess of thyroid hormone. In this same stage of sleep there is an outpouring of growth hormone which increases the synthesis of protein and of RNA, one of the two principal pattern-bearers in the cell. Thus the sound slow wave sleep appears essential to the speed of cell division, which leads to physical growth and repair of the body.

Dr. Oswald believes the dream period plays the same role in brain growth and renewal, particularly in the synthesis of proteins. At Edinburgh he and his colleagues studied the effects of heroin on their own sleep to measure restoration of brain function after drug usage. At first heroin reduced the amount of REM sleep. Later there was some recovery, but when the drug was stopped there was a large increase in the amount of dream sleep over a period of two months.

"It has seemed to me possible that the very large amount of dream sleep somehow assists in the synthetic processes of brain repair after injury," Dr. Oswald said. "The belief is consistent with the fact that in dream sleep, unlike slow wave sleep, the blood flow through the brain is increased far above waking levels. This suggests that it is a time of intense activity in the brain's living chemistry and to match this there is increased heat output by the brain. By contrast the blood flow through muscles falls by two-thirds during dream sleep."

To support the supposition of brain repair during dreaming, he points out that senile brain decay in the aged coincides with the fact that elderly people enjoy very little sleep in the dream state.[11]

So as Macbeth commented, sleep is truly the chief nourisher in life's feast and regardless of the number of hours we sleep each night, each part of each 90-minute cycle is essential to some part of the rhythm of life in body and brain, down to the infinitesimal clocks which determine the tempo of cell division. From sleep and other circadian research, Drs. Kleitman and Dement have suggested that the human being actually may live by "90 cycles." Brain waves follow a 90-second cycle. During sleep a person dreams in 90-minute cycles, and there are roughly 90 minutes of dreaming in nine hours of sleep. The 90-minute cycle, the two point out, still continues during the day although it is less noticeable.[12]

.. Sleep Learning

Is it enough for the inquisitive—and acquisitive—human being to understand the importance of sleep and how to use it better, or is it possible that fuller utilization might be made of this third of our life which we spend in recuperative oblivion? Thomas Edison thought sleep was a waste and trained himself to function with only a few hours each night (although historians report he napped a bit during the day).

One prospect which has persisted along the shadowy fringe of science and education for a half-century is the possibility that recorded messages repeated over and over could help us to learn while we sleep. Aldous Huxley's fictional prophecy Brave New World, published in 1932 but amazingly predictive of many social and biological changes visible today, spoke of a youngster who, while sleeping, heard a lecture by Bernard Shaw and the next morning was able to repeat it verbatim. Such miracles have not been revealed in the tons of literature which since then have purported to show the feasibility and success of sleep learning. While granting that research has been sporadic and insufficient, most Western scientists doubt that we can learn while we sleep.

On the other hand, sleep learning has been approved by the Ministry of Education in the Soviet Union for nearly a decade. Hypnopaedia, as the process is known in Russia, first was introduced on a mass basis in 1965 and 1966 when 200 residents of Dubna, a town near Moscow, were subjected to five-day-a-week instruction by radio in the English language. Good results were reported but statistical evidence is lacking.

Since then the Russians have developed sleep teaching to a relatively high art, and it is purportedly used in 180 educational institutes including secondary schools, polytechnicums, institutes of adult education, and colleges for academic, industrial, and military training. The technique is believed to be especially good in learning languages or any subject where a great deal of memorizing is required.

The Russians do not claim to teach pupils totally during their sleep. The training material first is introduced to the students just before bedtime. Then, as the student relaxes into sleep, the material is repeated again and again for thirty to forty minutes. No effort is made to continue the auditory lessons through the night, but the material is repeated again in the morning just as the students are beginning to awaken for the day.

Soviet teachers claim that using this method saves considerable time which would be required for memorizing and cramming during the day and speeds up the entire educational process. They also contend there is more long-term retention of information acquired by the hypnopaedic method than with conventionally taught courses.[13]

The Russians may have something, but most investigators agree more research is needed in the questionable arena of sleep learning. However, now that scientists are closing in factually upon the finer mechanisms of sleep, biological clocks, and brain function, hitherto unimagined possibilities may well emerge.

Beyond the rather crude method of trying to "force" learning through one of the five senses that is attempting to rest, some scientists see the possibility of locating the centers in the brain which control learning and memory. If this can be done, it may be possible to connect the human brain to a computer—bypassing the senses through which we normally absorb new experiences—and teach new information directly.

> Research indicates that sleep, which oc- curs in rhythmic 90- to 120-minute waves, including periodic dreams, is an ingrained cycle essential to the restoration of body and brain. We can make better use of this vital rhythm by understanding a number of things about it:
>
> 1. Dreams are normal in everyone and probably vital for restoring mental proc- esses.
> 2. Many insomniacs might cure them- selves by determining their own best hourly cycle for going to bed and rising in the morning.
> 3. Drugs and alcohol, including many sleeping potions, disturb the rhythms of sleep, adding that factor to other detri- ment they cause to mind and body.
> 4. It may someday become possible to use hours of sleep for increased learning.

CHAPTER 5

*The disorienting effects
of changes in time zones
or work shifts.
Easing transitions and adjustments
to keep equilibrium.*

When
the Body
Is Confused

AN AMERICAN BUSINESSMAN who planned a vacation
with his wife and children in Hawaii was called to Paris on
a business trip. He sent his family on to Hawaii and promised
to meet them there when his transactions were completed.

After ten days in Europe he was ready to return. He ate
breakfast in Paris before departure, and lunched on the plane.
When he arrived in New York it was 5 P.M. Paris time, but
noon at Kennedy Airport. So he ate lunch again before flying
on to San Francisco.

It was 6 P.M. when he arrived on the West Coast, but 9 P.M. New York time and 2 A.M. according to the Paris schedule he had been keeping. Despite the hops across a fourth of the world, which continued to turn beneath him, the businessman joined friends for a sumptuous dinner in San Francisco. His appetite was poor and he didn't enjoy the meal much.

Early the next morning he flew on to Hawaii. By then he was tired, irritable, mentally disoriented, and suffering acute indigestion. It required four days of rest for his biological clocks to adjust and catch up with him. And he was lucky; some persons would have needed as much as one day of catching up for each hour's difference in the time zones crossed.

In this technological world, artificially geared to split-second timing for most human activities, it seems there is no time to pay serious attention to proper operation of our internal clocks. Yet it was the jet airplane, product of modern technology, which revealed the importance of human circadian rhythms and the consequences of upsetting the cycles.

. . . Jet Fatigue

SHORTLY after the jet age dawned in the 1960s, people who flew off to a fast week of vacation or business in Europe or Hawaii began to complain of general discomfort and malaise which prevented them from enjoying themselves or working at a normal pace for several days after arrival. The symptoms were called jet fatigue. Doctors believed the vague disease was caused by the unusual and disorienting experience of flying while keyed up and excited, coupled with loss of sleep and irregular meals.

The tendency of a vacationer to ignore regular sleep and overindulge in food and drink, "making the most" of a holiday, obviously is physically upsetting. But it became apparent that more than tension, different foods, and excitement were at work. Dr. Charles A. Berry, who now is physician to the astronauts at NASA's Manned Spacecraft Center in Houston, speculated that part of the problem stemmed from the experience of flying *per se.*

"Any flight, whether a few hundred feet or a few hundred miles above the earth's surface places man in an environment for which he was not designed," Dr. Berry said in 1963. He pointed out that many civilian jet passengers undergo a subtle anxiety due to "loss of contact with the earth." [1]

This anxiety was part of jet fatigue, but we have since learned that most of the physical and mental disturbance results from the efforts of internal clocks to maintain a person's internal equilibrium on the "home" schedule while his body is flying off across new time zones at 500-plus miles an hour. As one doctor said, the conscious mind can understand it is in a new phase of the earth's rotation, but the body cannot.

In 1965 the Federal Aviation Agency set up an experiment to pin down some facts on what happens to a person's sleep-waking cycle, body temperature, and daily "tides" of adrenal hormones and other body secretions during jet flight. Four male volunteers, aged thirty to fifty-five, were taken to Oklahoma. For a week they were tested to establish base lines for pulse, blood pressure, breathing rate, urine flow and content, flicker-fusion time (a measure of how fast a light can flicker before it appears to merge into a steady beam), perspiration of the palms (an index of emotional tension), and rectal temperature.

After adjusting to Oklahoma time, the volunteers were flown west to Manila in the Philippines (a time change of ten hours) and tested again. Researchers found marked differences in physiological readings which required four days to readjust. The men then were flown back to Oklahoma. Disrhythmic symptoms seemed less on the return flight and disappeared faster.

The same experiment was conducted with volunteers flying eastward to Rome and return. Again they suffered differences in physical adjustment as well as in reaction times, decision-making ability, concentration, and attention. In Rome when the travelers were asked to punch a telegraph key on signal, their reaction times were almost twice as long as at home. Internal body temperatures took at least four days to shift to the new day-night cycle. Heart rate and water loss through perspiration took still longer to adjust.

These experiments confirmed that shifting a person rapidly across a number of time zones confuses the biological clocks, but it is not yet certain if north-south flights produce the same result. Relationship with the earth's 24-hour rotation was indicated by flying the volunteers from Washington, D.C. 5000 miles south to Santiago, Chile (only a one-hour time difference). This flight caused almost no physical or mental changes among the travelers.

"Shifting rapidly through time zones causes measurable disruptions in both physiological and psychological functions in humans," reported Dr. Sheldon Freud, Air Force psychologist who coordinated the test project. "It doesn't seem to matter whether people go to the east or to the west. Body functions are thrown out of kilter for three to five days—but apparently less after returning home. Mental adroitness is impaired for about 24 hours." [2]

The 1965 tests were done with trained scientific observers. In 1971 Trans World Airlines and Syntex Pharmaceutical Company of London conducted a new study using eight men and six women representative of average travelers. They flew from London to San Francisco, a nine-hour time difference, were tested for ten days, and then flew back to London, where they stayed another seven days. In this case, upsetting the biological clocks affected functioning of cells, glands, kidneys, liver, and the nervous system.

Some cycles shifted to new time schedules more rapidly than others. Body temperature took seven days to adjust when travelers flew west, and three to five days when they flew back east. It required at least three days for the biological clocks of all test subjects to adjust to new cycles.[3]

These field tests have shown how long it may take for the average person to feel "good" after throwing his biological cycles out of phase. Other studies established more definitively some of the things which happen beyond our sensory impressions.

In 1969 Dr. Elliot D. Weitzman, then at the Albert Einstein College of Medicine in the Bronx, New York, tested the effects of reversing the sleep-waking cycle of volunteers. Five young men lived for three weeks in a hospital ward. For seven and one-half days they were allowed to sleep at night for eight hours in an air-conditioned, quiet, dark room. They received a standard diet. Doctors measured brain waves, eye motion, and muscle tension. On the eighth evening the volunteers remained awake all night and then were allowed to sleep from 10 A.M. to 6 P.M. This schedule, 180 degrees out of standard phase, then was continued for two weeks.

Observers found, first, that all of the test subjects slept fewer hours after their schedule was reversed, but deep sleep followed normal waves. The major change occurred in dream sleep, which was considerably less immediately after the reversal and required the full two weeks of the second phase to adjust to near-normal levels. Throwing the sleep rhythm out of phase thus reduces dream sleep, which, we have seen, is vital to storage and sorting of memories and information and recuperation of the brain.[4]

People who seem to suffer most from time-shifting disrhythm are those who live well-ordered lives at home, then fly long distances on pleasure or business trips or otherwise shift their schedule abruptly. Although other than circadian factors may affect them (such as strange food and bad water), such travelers may suffer indigestion, wakefulness, fatigue, headache, sinusitis, bronchial conditions, and pains in the bones and joints.

. . . Easing the Transition Between Time Zones

SCIENTISTS point out that these symptoms are seldom serious for the person who allows enough time at his destination for his out-of-phase rhythms to adjust. This may vary from one to several days, depending upon the length and direction of flight. The International Civil Aviation Organization of Montreal set up a formula to help employees and travelers treat their internal clocks more kindly when traveling. The rules recommend that a person flying from Rome to New York, or vice versa, get one or two days rest for circadian adjustment after landing. If flying from Montreal or New York to Karachi, Pakistan, the traveler needs two and a half days of rest to catch up, say ICAO researchers.

For most earthbound creatures and occasional travelers, disturbing the internal clocks may mean no more than a ruined vacation. It is more serious for people who must remain alert for important action or decisions.

"People flying long distances should take this adjustment into account," said Dr. C. C. Gullett, director of medical services for TWA, "especially diplomats, government leaders, military officials, businessmen and others who try to make important decisions after an extended flight." [5]

Drs. Stanley R. Mohler, J. Robert Dille, and H. L. Gibbons of the Federal Aviation Agency have suggested that many vacations can be made more pleasant and business trips more successful by planning, pacing of activities, and rest. Several days before a trip is scheduled, the doctors advise, a person should shift his hours of eating and sleeping gradually toward the time schedule of his destination. If this is impossible, he should at least be well-rested before starting.

The journey should be timed for arrival in the evening so that a full night's sleep is first on the agenda. If a person does not feel like sleeping then, he should induce physical fatigue by walking or other exercise. The doctors also warn against heavy eating and drinking while the body and mind are out of phase with local time, because this intensifies jet fatigue. The most important point is to allow at least a day after arrival before participating in demanding activities such as sight-seeing or business appointments which require physical and mental alertness. As another doctor commented, a jet tour offering ten countries in seventeen days is an invitation to misery.

"Above all, take it easy," Dr. Mohler advised, "because several days are required for complete resynchronization of the biological rhythms after flights through several time zones." [6]

Somewhat contrary advice was issued in 1971 by Karl Klein of the Institute for Flight Medicine, Bonn-Bad-Godesberg, West Germany. He suggested that travelers should move into a new routine quickly after a flight, rather than staying in their hotel rooms, so that their rhythms would adjust more quickly. Klein told the International Society for the Study of Biological Rhythms that flying east is harder on body rhythms than flying west because a person loses more sleep flying east. Individuals recuperate at different rates. Some may recover within one or two days, others might take a week or two. Although temperature cycles usually are back to normal within a week, isolation of subjects after flights has kept their temperature from rephasing for fifteen days, according to Klein.[7]

*

FIGHTING FLIGHT LAG: Some Guidelines for Escaping the Worst Effects of Jet-Age Travel

Eastbound Flights

1. Where possible, make such flights in daylight hours, leaving as early as possible—earlier as the distance increases.

2. To achieve best sleep conditions, log the following times:
   Departure: _____
   Arrival: _____

   In order to achieve my normal _____ hours nightly sleep, I must retire at _____ P.M. (local time) and arise at _____ A.M.

Westbound Flights

1. Where possible, make such flights late in the day, arriving as close to your normal retiring hour (local time) as possible.

2. On the day, or days, prior to departure, try to retire one hour later each day for every time zone you will fly through. (E.g.: If flying from New York to Los Angeles—a three-hour difference—retire one hour later three days before departing, two hours later two days before departing, three hours

later one day before departing.) In this way your body adjusts more readily to the local time zone.

Number of time zones in flight _____

Number of days to retire later _____

First day: adjusted bedtime _____ P.M.

Second day: adjusted bedtime _____ P.M.

Third day: adjusted bedtime _____ P.M.

_____ day: adjusted bedtime _____ P.M.

_____ day: adjusted bedtime _____ P.M.

General hints: North-south flights need no special preparations unless there is a considerable east-west factor involved as well.

Try to break up long trips so that you fly no more than eight to ten hours on any one leg of the trip, with a one-day stopover between each leg if possible.

If you cannot have advance warning of your flight to arrange it as per the above directions, then:

1. If *traveling eastward* at night, stay awake throughout the next day. Do not take a nap in the morning upon arrival or during the day. Then retire the first night as early as possible, getting a full night's sleep, which will bring you back close to normal.

2. If *traveling westward,* eat very lightly throughout the day. You may find yourself having more meals than usual. Try to retire as though you were still in your own time zone, even though this may mean going to bed several hours "earlier" in terms of the local time zone.

Respect for circadian rhythm is essential for airline pilots, who may cross many time zones in several flights within a week. If a pilot is not given sufficient time to recuperate, disorientation might alter his ability to make quick decisions and thus endanger his passengers. Many air crewmen, when on duty, keep their watches set to "home" time and continue eating and sleeping by that schedule regardless of the time of day or night in their new location. In this way their body clocks remain partially adjusted while they are away and require less readjustment on the return flight.

Air crew members who do not pay special attention to their personal rhythms report a wide variety of symptoms. One TWA pilot described the jet syndrome as a gradual progression from headaches to burning or unfocused eyes, stomach problems and appetite loss, shortness of breath, sweating and, sometimes, nightmares. Stewardesses also have reported stomach problems, insomnia, elusive mental confusion, and menstrual irregularities.

The Russian Ministry of Civil Aviation has studied crews on nine-hour flights from Moscow to Khabarovsk, near China. They found changes in brain wave readings and other physiological differences, and now encourage scheduling that maintains a flight crew on a stable work-rest cycle. It is reported that Soviet pilots who fly to Cuba are boarded at a special hotel which runs on Moscow time.[8]

Other flight crews traveling from Amsterdam to Anchorage, Alaska and Tokyo found it took six days before urinary excretions of electrolytes adapted to local rhythm. Drs. Edmund B. Flink and Richard T. Doe at the University of Minnesota examined adrenal hormones and electrolytes in urine after east-west flights. One investigator flew from Minneapolis to Tokyo and on to Seoul, Korea, taking urine samples every three hours. It required nine to eleven days before his excretionary rhythms adjusted to Korean time, an adaptation rate of about one day for each hourly time zone crossed in flight.

A study of 150 pilots on the France-South Pacific route revealed that more than 70 percent of jet plane crews had difficulty going to sleep after landing in Papeete, Tahiti. Quality of sleep was altered and the pilots woke up frequently. After ten to fifteen days in the Pacific, the aircrews adapted to local time but then experienced similar disturbances when they returned to Paris. About 20 percent of the pilots resorted to drugs to help them sleep.[9]

These finer points of circadian phase shift and their consequences still are being defined, but for the moment it is enough to know the causes of jet fatigue and alert oneself to compensate for it when traveling.

In 1966 Lowell Thomas, world traveler and author, vividly described how his upset time clocks caught up with him (or rather, failed to catch up). In a period of several months he flew from New York to the South Pole and home by way of New Zealand, Australia, India, Iran, and Europe. Then he went on an expedition to New Guinea and flew home, circling the globe again. After that he traveled to Australia, Afghanistan, and Moscow, then to New York and finally to Detroit, where he was scheduled to deliver a series of lectures. During those months he had crossed all twenty-four zones of the world at least twice and some of them more. At one point he fainted on a train in Austria, and then lost consciousness a second time in a London hotel lounge. When he reached Detroit he was seriously ill and was taken to a hospital with what doctors first thought was a heart attack. After adequate rest, Thomas and his doctors agreed that his body simply had rebelled against the frequent and radical disruption of his internal rhythms.[10]

Throwing the clocks out of phase also may be more serious than a few days of discomfort or illness. As early as the mid-1960s some flight physicians detected preliminary signs of premature aging among pilots on regular east-west flight runs.

That sign, though not yet substantiated, has been supported in animal experiments by Dr. Halberg at Minnesota, who discovered it is possible to shorten the life-span of mice by inverting their light-dark schedule. In one experiment, two groups of mice were placed on a schedule of twelve hours darkness and twelve of light. Later one of the groups was subjected once each week to an inversion of lighting schedule, which would be comparable to a man flying halfway around the world. (That is an easier schedule than that of a transatlantic pilot, who might have to make two such phase shifts in a week.) The mice which had kept a standard routine lived an average of 94.5 weeks. The mice subjected to light-time reversal lived an average of only 88.6 weeks, a decrease of 6 percent in life-span.

In view of the mounting evidence of biorhythmic upsets, most nations and airlines now assure extended rest periods for flight crews between east-west assignments. Also, since rapid travel has become a way of life for a fairly large segment of the population, some doctors have begun trying to find ways to accelerate resynchronization of the internal clocks. Dr. F. Gerritzen in the Netherlands attempted to shorten the adaptation time on a test flight from Amsterdam to Tokyo by exposing seven volunteers to an inverse lighting schedule. The passengers, however, did not adapt according to expectation. Their excretion of certain hormones apparently was increased out of phase by the stress of the flight. The hormone rhythm, it appears, is one of the main facets of phase-shifting which causes mental and physical ill effects. An attack at that end of the problem is the motive for the study undertaken by Syntex Corporation with Trans World Airlines.[11]

. . How Changing Shifts
Affects Job Performance

THE importance of circadian rhythms is by no means limited to people who fly. The same disrhythms and discomfort disturb those, for example, who are shifted abruptly from day to night work or vice versa. Such people blame their unease and lower efficiency on poor sleep and are hardly aware of the gradual readjustment of their internal clocks during succeeding days or weeks. Most of us become aware of our biological clocks only when they are thrown out of adjustment, but the clocks are operating in all of us all of the time.

Long before the importance of circadian rhythms had been established, work efficiency surveys indicated discrepancies in performance among people working other than regular daytime hours. Russian scientists found more than normal errors and accidents, and increased ill health, among nightworkers on the Moscow subway system. These were blamed upon poor daytime rest, but they apparently were due more to phase shift problems.

A Swedish study indicated that meter readers in a gas works made most errors on night shift, fewer during afternoon, and least during the morning shift. In 1949 a U.S. investigator timed telephone operators as they answered incoming calls and found they were at their slowest around 4 A.M. In 1950 a Hamburg, Germany researcher found that industrial accidents occurred most frequently between 10 P.M. and 2 A.M.

The detriment to health and efficiency of rotating work shifts was further indicated in 1957 when a high incidence of ulcers was found among shift workers. Air traffic controllers usually rotate shifts, every few days at some airports, to prevent the same few men from handling the full-time burden of peak takeoffs and landings. These may occur every few seconds at major airports between 7 and 9 A.M. and 4 and

8 P.M. However, the men who seem to adapt to the rotating schedules also often suffer ulcers and hypertension.

A survey of more than 1000 industrial workers in the Rhone Valley revealed that 45 percent of the workers could not tolerate a work schedule which rotated every seven days, while 34 percent could not adjust to a two-day rotation. Body temperature rhythms did not adapt to either schedule. Even though a person's sensory adaptability may enable him to function with habits changed periodically, throwing the internal clocks upside down may cause long-term adverse effects on health.

Searching for these relationships to efficiency and health, British and American scientists observed men in the classic rotating watches of the Navy. This carry-over from medieval times, when men worked so hard they could seldom tolerate eight hours of it, breaks the day into watches of four hours each. In an average situation, a man would put in eight hours of work each 24-hour day, and in seventy-two hours would have worked each of six different watches. Dr. Kleitman, in 1948, took temperatures of submarine sailors who were on a three-watch system, essentially living a 12-hour (instead of 24-hour) day and never sleeping for eight hours at a stretch. Despite their outward adaptation to such a schedule, temperature readings showed their bodies still preserved a 24-hour day and they were most alert at noon. Test results showed this watch system would have left many of the men only partially alert in case of an emergency.[12]

Drs. N. I. Andreshyuk and A. A. Vesolova tested men in Russia on a standard daily routine for fifteen days; then, after a month's rest, on an 18-hour day. After another month of normal living, the volunteers were tested on a schedule of six hours of work and six of rest. The 18-hour day was most detrimental to performance. It left the volunteers drowsy, restless, and emotionally tense. They performed exercises abnormally fast but with poor coordination. One man lost muscle power; all performed inaccurately on tests.

Individual differences in response to time disruption lead to the growing belief that in a highly complex society, where pilots, hospital personnel, and others must live and work under phase shifts and unusual schedules, it may become imperative to develop tests that will screen people for time stability.[13] Such a procedure, however, must wait until enough data has been accumulated to devise a test that would be meaningful in predicting a time-stable person and one who is not. In the meantime, there are some measures which can be taken by a person in being shifted from day to night duty or vice versa.

Two or more free days, such as a long weekend, should precede such a shift. During this time, the individual should begin conditioning himself to eating and sleeping according to his new work schedule. In this way part of his adaptation will have taken place before he begins work and is expected to be at the peak of his mental and physical powers. If such adjustment time is not allowed, then industrial and other supervisors should understand that a certain level of inefficiency is to be expected until adjustment takes place. This can be crucial in jobs where split-second timing in action and decision is necessary.

Dr. Klein in Germany found great individual differences among pilots tested on a flight simulator around the clock for several days, then flown to the United States and returned. Each man sat at a panel simulating flight of a supersonic plane. They were tested against sudden winds, near accidents and other flight emergencies. One maneuver, simulating a flight adjustment, required an average of 53.4 seconds to perform in the afternoon but 103.3 seconds at 3 A.M. Individual performance changed as much as 50 percent from day to night.[14]

... Studying Circadian Rhythms in Astronauts

ONE of the pioneering sectors of circadian rhythm research seeks to learn the effect of disturbed biological clocks during manned spaceflight. Although well-trained astronauts and cosmonauts have weathered many days and nights in orbit and several on the moon, it is not yet clear what role circadian upset may play in long-term flight when the earth's 24-hour day is traded for the ninety minutes of day and night in a typical earth orbit. Measurement of these effects is difficult because weightlessness, which man is experiencing in long duration for the first time in history, also causes many physiological changes. It will require many more hours of closely monitored space flight to delineate between the effects of weightlessness and circadian phase shift, but it already is apparent that sleep is profoundly disturbed.

Frank Borman slept badly early in his 14-day Gemini orbital flight in 1965. On his first night he alternated between light sleep and arousal, receiving little slow wave sleep. By the second night he had adapted somewhat and received some slow wave sleep to compensate for his loss the night before.

In deference to the need for keeping step with earth schedules, American astronauts set their chronometers to Cape Kennedy time, the point of their launch. However, the astronauts have operated by several different sleep-rest cycles, ranging from a standard daily schedule to alternation between four hours duty and four of sleep. While establishing work-rest schedules for the cosmonauts, Soviet scientists tested for mental and muscular efficiency, balance, pulse pressure, and other heart, blood, and urinary functions. Stressing the importance of normal circadian schedules, the cosmonauts have been maintained on their earthly sleep cycles while in orbit.

One factor contributing to sleeping difficulty in space is disorientation of the vestibular system, which governs the sense of balance. Without gravity, this system has no point of reference, so weightlessness may play a role in upset biological rhythms. Dr. Berry at NASA has described how sleep-hungry astronauts try to orient themselves by wrapping an arm or leg around something while free-floating in space. The same difficulty was experienced by the Russians. In 1961 when Titov spent twenty-five hours in orbit, he fell asleep at his usual bedtime but awakened early, disoriented by the sight of his arms and hands dangling motionless in midair.

"The sight was incredible," he said. "I pulled my arms down and folded them across my chest. Everything was fine—until I relaxed. My arms floated away from me again as quickly as the conscious pressure of my muscles relaxed and I passed into sleep. Two or three attempts at sleep in this manner proved fruitless. Finally, I tucked my arms beneath a belt. In seconds, I was again soundly asleep." [15]

Manned space flight, as astronauts are removed from earth for longer periods of time, eventually may resolve the question of whether biological clocks are internal and independent, or if their operation depends upon a continual flow of time cues from the environment. If the latter is predominantly true, then during long expeditions in space a human being might adjust gradually but permanently to new forces encountered, whatever they may be. If the clocks are deeply ingrained and resistant to change, however, then the longer a man stays in space, out of phase with his natural circadian periodicity, the more his mental and physical performance might deteriorate.

The longest American manned mission in earth orbit was the 14-day flight of Gemini VII, which was designed primarily to determine if men could live and work long enough in space to complete a mission to and from the moon. While proving that feasibility, Gemini VII also provided valuable information on the persistence of circadian rhythms.

Drs. Kenneth N. Beers and John A. Rummel of NASA analyzed continuous electrocardiograms with readings of heart function telemetered to earth while the flight was in progress. The investigators found that the pilot and command pilot varied to some degree in both quantity and quality of sleep. Despite the fact that the men were removed from earth's gravity in a 90-minute revolution around the home planet, most heart functions held steadily to 24-hour cycles. Most interesting was the fact that the astronauts' internal functions followed a day varying between 23.3 and 23.6 hours. This periodicity coincided almost exactly with the operational shortening of the flight "day" to compensate for orbital precession of the spacecraft caused by rotation of the earth beneath it.[16]

Whereas the main thrust of American space flight has been toward exploration of the moon, the Soviet program has aimed more specifically at testing man's resilience during long periods in the alien environment of orbital flight. Soyuz 9 in 1970 was in space for eighteen days. Upon return the cosmonauts suffered lower blood vessel tension and for several days were unable to stand erect without help. Heart function also was impaired, posing the danger of heart failure during the sharp transition from weightlessness to high multiples of gravity when a spacecraft re-enters the earth's atmosphere. The mission of Soyuz 11 in 1971 extended manned weightless flight for twenty-four days. The mission undoubtedly would have added valuable data on the factors of biological rhythms as well as weightlessness. Unfortunately, it ended tragically when the cosmonauts died by malfunction of their spacecraft during re-entry.[17]

Now that the moon has yielded some of its geological secrets to Apollo explorers, the American space program also is turning toward long-term living and working in earth orbit. Beginning in 1973, the United States will launch a series of

Skylabs in which astronauts and scientists will live and work for periods of twenty-eight to fifty-six days.

One of the tools which Skylab astronauts will carry with them is a sleep analyzer designed by Dr. James D. Frost of Baylor University and developed to compact portability by SCI Electronics of Houston, Texas. This machine combines and evaluates electroencephalograms for reading brain waves and electroocculogram signals to measure eye movements. Analysis of these functions will help scientists determine the rhythms and quality of sleep while the astronauts are in orbit.[18]

This machine is an example of the extensive list of benefits which have derived from the nation's space program to help solve more earthly problems. The sleep analyzer may be used to study sleep patterns of air traffic controllers and pilots after periods of stressful duty. It may help also to evaluate the sleep patterns of drug addicts and in identifying harmful side effects of new drugs. Such tools are essential to scientists as they trace the functioning of our biological clocks, the consequences of disrupting normal phases, and conditions for obtaining highest efficiency from our internal rhythms.

Throwing biological clocks out of phase, as in jet travel across time zones or changing time shifts in work, can cause physical and mental upset ranging from discomfort to illness. Awareness of the need to give internal rhythms time to adjust to new circumstances will enable a traveler or person changing work shifts to take steps to make the adjustment without difficulty.

Some general rules:

1. When planning a long-distance trip,

such as to Europe or the Orient, begin to shift eating and sleeping habits toward the time schedule of the new destination.

2. Plan arrival in the evening so that a full night's sleep is first on the agenda.

3. Allow at least a day before taking part in strenuous physical or mental activity, such as sight-seeing or important business conferences.

4. To achieve maximum efficiency and comfort for people assigned to work shift changes, supervisors should allow employees time to adjust to their new schedule. This is most important for work demanding high skill and alertness, such as flying and air traffic control.

One of the benefits likely to derive from future manned flights in space will be determination of man's ability to adjust to phase shifts and disorientation in time.

CHAPTER 6

*The cycles in
emotional ups and downs.
Glands and beats—charting
for peaks of good times and bad,
allowing for them.*

Human Moods: Unaccountable or Predictable?

A TRAVELING SALESMAN, through years of experience and self-observation, discovered that his moods changed on a 48-hour period. One day he would be so depressed, irritated, and lethargic that he could barely force himself to contact prospective customers. The next day he would feel elated and energetic.

This man learned to make calls on clients only every other day, assigning routine work to his "off" days, and found that the pattern of his entire working life improved.[1]

In Germany a thirty-five-year-old mother of two children suffered from recurring periods of lassitude and depression coupled with anxiety and fear concerning the health of her children. Her energy increased with the beginning of each menstrual cycle, but then she suffered psychotic episodes in which she was certain everyone in the world was about to die. Electroshock and treatment with drugs failed to alter the intensity or rhythm of her mood changes. Finally, therapy with the hormone estrogen, timed to the proper phases of her specific biological rhythm, proved successful in erasing her feelings of incompetence, weariness, and despair.[2]

Dr. Bruce Quarrington, a psychologist at the Clarke Institute of Psychiatry in Toronto, Canada studied mood changes in 200 volunteers. He had both men and women observe their feelings three times a day over a period of thirteen weeks. Many showed predictable mood patterns which were not related to changing events in their lives.

Fifteen percent of the group—both men and women—showed a weekly up-and-down mood rhythm. The high point was Friday and the low point Tuesday, but apparently not related to the average person's anticipation of a weekend free of work. A few individual cycles were longer. In most of the women this seemed related to the ebb and flow of their menstrual periods.[3]

In New York City, teachers at a Bronx school have found success with an experimental program using awareness of rhythms in cultivating a young pupil's capacities to learn.

"This program is very rhythm-oriented," said Mrs. Elton Warren, music supervisor of the school. "There's a rhythm to life. Each person has to find his own beat. We encourage each child to do this in the classroom and show it; then we take it and use it. It's an approach to developing the whole child. We start with him, his body, the way he feels about himself, the way he learns. Each child learns differently." From rhythm in music, the program has helped to show the tempo by which an individual may learn many things.[4]

The salesman, an anxious mother, children in a Bronx school —all are random examples of broadening awareness that brain and body are inextricably linked in the multiple cycles of biological timekeeping. The time of day, week, month, or year when a person experiences events may exert influence upon his depth and swing of emotions, ability to learn, and recording and recall of memories.

Scientific exploration of the brain is just now beginning to unlock knowledge of the seats of emotion, learning, and memory, but all three appear to be keyed with periodic function of the endocrine system, particularly the adrenal glands. Adrenal hormones have a bearing upon emotion and vitality, and response to various events often depends upon the phase of the circadian cycle in which the events take place. Emotional conditioning—along with vigilance, sensory acuity, and drug response—is tempered according to the biological time of day in an individual. Such fluctuating rhythm may help to explain how stressful situations affect a person and sometimes lead to emotional instability and illness.

An example is "postnatal blues," depression suffered by many women two to four days after giving birth to a child. In about one case in a thousand, the blues turn to psychosis so severe that the woman must be hospitalized. One twenty-seven-year-old woman, known for good mental balance, became pregnant after two years of marriage. She and her husband both were pleased at the prospect of having a child, and she enjoyed a normal delivery.

Late in the third day after the child's birth, the woman began to feel depressed and worried. The symptoms did not go away when she went home from the hospital. Her mother was there to help and noticed her daughter was behaving oddly.

"It was just like sunshine and cloud," the baby's grand-mother related. "Part of the day she'd be fine. Then she'd be dressing the baby and all of a sudden say, 'Well, that's enough of that!' and walk away leaving the baby lying alone on the table."

Later the young mother began having hallucinations and was hospitalized. A doctor diagnosed her condition as postpartum psychosis. She improved under psychiatric care and electro-shock for several months but seemed to have a relapse just before each menstrual period. Two years after her recovery, the woman had another child. The symptoms repeated almost exactly, including the necessity for three more months in a mental hospital. After her recovery that time, she decided to have no more children.[5]

Because the postnatal blues occur at predictable times, and in women who have no history of emotional disturbance, many doctors believe the condition is caused by a chemical or hormone imbalance triggered by an internal timing mechanism responding to the stress of childbirth.

. . . Long Mood Cycles

MOST emotional swings related to biological clocks are not as obvious as those in the menstrual cycle or postnatal blues. In most of us, long undulations in emotion and psychological change may occur unnoticed. Such were the previously men-tioned findings of the late Dr. Hersey of Pennsylvania, who dis-covered long mood cycles among industrial workers.

Individual charts showed that emotional tone varied within each day, but longer trends also were detected. One congenial sixty-year-old man, who claimed that he never changed, actu-ally had a 9-week mood cycle with a decline so gradual that he did not realize, in his low period, that he was no longer joking with his coworkers while tending to withdraw and criticize his superiors. A twenty-two-year-old man displayed a 4½-week cycle of emotion as regular as the average woman's menstrual period. During his low cycles this young man was indifferent, apathetic at work and at home, and temporarily abandoned his art work hobby. Another more temperamental person, with a cycle of 4½ to 6½ weeks, was irritable and magnified crises out of proportion during his low period. A fourth man, with a 5- to 6-week cycle, had periods of great vigor and almost frantic energy when he felt confident and outgoing. In his low periods he found work a burden, slept more, and spent long periods sitting quietly. He weighed less and slept less in his exalted periods.

Adolescent boys sometimes suffer monthly psychotic episodes, and less obvious undulations of mood may underlie what all average men and women consider to be their normal state of being. Unfortunately, most healthy people never have occasion to chart their mood swings accurately enough to recognize their own internal rhythms. Most of us "muddle through" our habitual work during the low phase of emotional cycles, assigning our moods to other environmental factors and life changes which obviously do have a great influence upon how we feel and function from day to day. As a result, the so-called normal person is never observed medically over a long enough time to define his periodicities of emotion, learning, and memory.

In one respect, this may be an advantage in our frenzied society, which already tends to stimulate morbid anxiety about personal health. On the other hand, failure to understand our natural cycles can lead to unnecessary concern when we are at low ebb. If we learned to recognize our own periodic swings in emotion and mental efficiency, this knowledge would eliminate needless worry and might improve achievement in both the high and low periods. It also might reduce the number of times we reach for the pill bottle to provide a temporary lift.

At the present stage of medical practice, emotional cycles go unrecognized until the shifts are so severe that medication is called for. Then drugs are applied to ease the condition, masking the rhythms and possibly changing the cycle of the rhythms themselves. A true base line for mood and behavior might reveal individual cycles of several frequencies—monthly, seasonal, and annual as well as the daily swings of circadian periodicity. At the moment, there is no way to estimate how much a person's sense of well-being might be enhanced by living according to his own particular beat in the harmonics of the universe.

. . . Sunspots and Emotional Outbursts

THE hypothesis that mental processes may indeed be synchronized with the harmonics of our solar system is supported by studying the erratic behavior of the mentally ill. In many hospitals attendants have found that aggressive patients become intensely active and agitated at periodic intervals. They may change unaccountably from passive behavior to hostility and violence. For centuries such outbursts have been considered idiosyncrasies of individuals whose violent actions may touch off mass reaction among other patients. Evidence now indicates such behavior may occur according to predictable rhythm.

At Douglas Hospital in Montreal, doctors studied periodic outbursts and other patient behavior around the clock. They tried to correlate increased aggression with changes in menu, medication, visiting days, and even change of staff members on duty. Barometric pressure, temperature, humidity, and other environmental factors were compared with the periods of recurring agitation, but none of these seemed to account for it. Finally, Dr. Heinz Lehmann compared his hospital data with information from the U.S. Space Disturbance Forecast Center in Boulder, Colorado. To his surprise, he found a correlation between excitement in the mental ward and the time of occurrence of solar flares, the well-known sun spots which accompany geomagnetic disturbances on the surface of our ruling star.[6]

Investigators are hesitant to accept circumstantial evidence without corroborating data, but such a relationship is not impossible. Solar flares are bursts of gaseous material which send a flood of high-energy particles through the solar system. These bursts of solar wind influence the ionosphere and alter the magnetic fields on earth sufficiently to distort radio communications and affect a compass needle. It is possible that the brain, or at least a deranged one, may be sensitive to such influences and our internal clocks respond to environmental factors beyond the range of our standard senses.

Earth dwellers are subject to a multitude of environmental forces, some stemming from genetic heritage, others ranging out to the flood of solar radiation which is ultimately responsible for all sources of energy. Some influences, including man-made stress, are more dominant than others in altering mood and behavior. It is logical that the higher the mental development of a species, the more sensitive that species may be to subtle changes in the universe. It is paradoxical that man's mental development, which has made him the most adaptable and dominant creature, has at the same time separated him from instincts which categorize the behavior of lower organisms and confused statistics by making each person an individual. This is why it is difficult to define "normal" and "average" when dealing with mental and emotional responses and their cyclic ebb and flow.

. . . Linkage of Adrenal Hormones and Emotion

ASIDE from the reactions of clearly abnormal people, it appears the average person's emotional rhythm may be linked to cycles of hormone secretion from the adrenal glands. The rhythmic fluctuation of adrenal steroids in human urine was first detected in 1948. During the following decade, Dr. Halberg at Minnesota discovered through numerous studies that a healthy man on a regular sleep schedule has a regular rise and fall of adrenal hormone levels in blood and urine, a cycle that occurs every twenty-four hours. The precise hours of peak and low ebb seem to depend upon individual work and rest schedules, as well as social factors. The rhythm of hormones shifts gradually when a person changes schedule, but in a healthy person it holds to a 24-hour period.

Persistence of the adrenal cycle was first noticed in three mental patients who were receiving electroshock therapy. So deep was their psychoses that these patients lost all concept' of time, had no memory, and were unable to speak. Under such conditions, researchers expected to find the adrenal cycle deeply disturbed. Instead they discovered the cycles still were continuing with circadian periodicity.

This predictable steadiness is important for two reasons. First, it is known that abnormal adrenal hormone levels may be a factor in mental depression. Second, since most individual adrenal cycles are not precisely twenty-four hours in length, it is possible for this important clock to slip out of phase with other rhythms over a period of time. It would require hormone measurements at the same time day after day for several months to learn whether an individual's cycle was advancing or slipping behind a few minutes each day. Over a period of several weeks, such a shift out of phase with daily living and other internal clocks could account for a monthly, or longer, cycle of elation and depression.

The disadvantage of being desynchronized from one's normal environment are obvious, as we have seen in the upsets caused by jet travel across time zones. Mental depression, from throwing mind and body out of phase with the adrenal cycle, may be one of the results. When coupled with emotional disturbances from other sources, upsetting the rhythm may be the trigger for disaster, including suicide.

Recognition of such possibilities led Dr. William Bunney, Jr. at the National Institute of Mental Health in Bethesda, Maryland to relate adrenal hormone levels with inner stress suffered by mentally depressed patients. Understanding such relationship may help to identify potential suicide victims.

Investigators at the Eastern Pennsylvania Psychiatric Institute in Philadelphia compared adrenal hormone levels at different hours among healthy and depressed persons. They matched twenty healthy women and twenty healthy men with ten men and women who were suffering severe depression, anxiety, agitation, or anger at the psychotic level. The tests revealed a daily variation in mood paralleling the rhythm of adrenal hormones in the blood. The most noticeable difference was between sexes. Emotionally disturbed men showed a higher amplitude of change in their hormone rhythm than both normal men and women. Overall, the daily cycles of adrenal hormones were stable among healthy persons but variable among the emotionally disturbed.

These cycles of depression as they relate to adrenal hormones may be partially due to dissociation with other rhythms rather than just a change in the level of a specific substance. Thus a comparison of several internal functions with behavior over a period of time may become a fruitful way of studying and treating depression. Doctors must pay attention to changing mood and behavior as well as the cyclic flow of hormones and other body chemicals. One pioneering technique has been developed at the Institute of Learning in Hartford, Connecticut, where detailed nursing notes are compared with physiological measurements and analyzed by computer.

The nursing notes are based upon a detailed checklist of traits, actions, and descriptions providing a comprehensive evaluation of a patient every twelve hours. The nurse fills out a printed form with 215 descriptive statements. These include items on care of the patient's appearance, whether he groomed himself, and whether he seemed gloomy, irritable, angry, tearful, preoccupied, or cheerful. The list examines his sleep, whether or not he took his medicine regularly, and if he attended classes. The forms then are analyzed by computer, showing a patient's profile as it changes over periods of days or weeks.

Analysis of one patient revealed that as his depression improved, his behavior fluctuated in a three-day rhythm. Also, there were distinct phases in his improvement. First, his thinking became better organized although he remained anxious and depressed. His anxiety decreased next, and finally his depression went away. Thus careful and frequent examination coupled with computer analysis of fluctuating rhythms may enhance the ability of doctors to deal with nebulous mental disorders.

The importance of adrenal hormones to the up-and-down waves of physical and mental processes has tempted scientists to search for the central controller, or master biological clock, in the adrenal glands. These also secrete from their core epinephrine (adrenaline) and norepinephrine (noradrenaline). If the adrenals are isolated from all other organs, they continue to secrete hormones in rhythmic cycles. This does not mean, however, that we are likely to find a tiny organic clock ticking away the seconds in some remote fold of gland or flesh. The rhythm exists in single cells, and a circadian beat persists in human tissue removed and nurtured in laboratory cultures.

In any case, the track of the adrenal rhythm extends at least one step farther to the pituitary gland, which releases ACTH (adrenocorticotrophic hormone) and stimulates adrenal secretions. One study showed that patients with severe brain damage had no obvious circadian rhythm of adrenal hormones in the blood. This suggests that the brain controls the pituitary gland's secretion of ACTH, which in turn regulates adrenal steroid hormore production. Other tests indicate the rhythmic clocks which influence thinking and the transmission of nerve impulses may not exist in a central point, but may operate at several different localities in the nervous system.[7]

. . Critical Periods in Learning and Memory

WHILE that biochemical problem continues to puzzle many researchers, psychologists have begun to define critical cyclic periods in learning and memory. Because of the difficulty in manipulating human beings, the basic statistical clues must come from animal studies. These have progressed in complexity and sophistication since the days of testing rats in a maze or Pavlov's conditioned reflex experiments with dogs.

Some experiments involve training animals to perform a task and then applying drugs or other stimulus to see if the creatures remember the task. These techniques may shed light upon some of the strange amnesias people suffer after traumatic experiences such as strokes or accidents. They also may show how the human nervous system begins to malfunction under stress and frustration, and why some memories are retained while others are forgotten.

Scientists have discovered there are critical periods in the circadian rhythm for learning fear and retaining memory of it. One of the most respected scholars in this area is Dr. Charles F. Stroebel of the Institute of Living in Hartford, who began asking a decade ago how the biological time of day might influence learning. His continuing research has revealed a number of intriguing answers.

One training procedure, with a rat or monkey, is to induce the animal to press a bar at a certain time to receive rewards, such as food and water. When the animal has learned this lesson, then it is possible to see how drugs, shock, or other factors affect his bar-pressing habit. A stimulus such as light or sound may be associated with a reward, or with an unpleasant experience such as electric shock.

The Institute's first study of biological rhythms and their impact upon emotional learning utilized rats trained, under a rigid schedule of darkness and light, to press a bar regularly for a reward of water. After conditioning, as the rats pressed the bar for their reward, they also would hear clicking sounds followed by an electric shock. After nine training sessions, the typical rat had learned that the clicks and shocks were unavoidable. When it heard the clicks it would become fearful and stop pressing the bar. Soon the animals developed signs of definite anxiety.

This test schedule was patterned after the way in which human beings also are believed to learn anxiety. A parent may reprimand a child before spanking him, and thereafter the child becomes anxious when he hears a certain tone of voice— even years later when the punishment itself has been long gone. Psychologists find that, in most people, such an anxiety eventually dies out or is *extinguished.* Helping this to happen is one of the major goals of psychotherapy. In a mentally disturbed person, however, the anxiety syndrome may persist illogically long after the source of a fear has passed.

With the laboratory rats, undoing or extinguishing the fear response was accomplished by allowing them to rest several days and then confronting them with clicks *without* shocks when they pressed the reward bar. The test animals were divided into three groups, each subjected to a different pattern of light and darkness.

One of Dr. Stroebel's first findings was that fear was strongest and most difficult to extinguish at the time in its biological day when the subject had been trained. If a rat learned fear at 8 A.M., his fear responses were stronger at 8 A.M. Beyond this, those which had been conditioned and learned fear at the same hour each day took longer than others to erase fear from their memory. Also, the fear was extinguished successfully *only* if they were retrained at the same biological time of day as their anxieties were learned. Moreover, the studies indicated that if fear is taught over a random time schedule rather than at regular and precise hours, then anxieties can be removed only under similar random conditions.

This suggests that if a person's anxieties or fears are acquired at random times of day and night over many years, then psychotherapy to remove these anxieties also should follow a random pattern rather than through conditioning at the same hour each day.

Linking depression with the adrenal hormone cycle, Dr. Stroebel found through blood analysis that the greatest susceptibility to fear-learning occurred when blood hormone (corticosterone) levels were high. The anxiety studies also gave a clue which may help to explain some psychosomatic illnesses in humans. Long after the experiments were over, Dr. Stroebel noticed that some of his test subjects still were suffering recurrent hyperventilation and high blood acid levels. These symptoms occurred almost exactly at the time of day when the subjects formerly had undergone trials in unavoidable fear.

The cyclic recurrence of high acid levels, after emotional stress had been removed, sent researchers toward new investigations because the acid-base balance of the blood is a delicate equilibrium which affects the body in critical ways. Dr. Stroebel turned to monkeys since their day-night patterns and physiology are more similar to humans.

Under training, the monkeys soon learned to dread the electric shocks and the clicks which preceded them, showing signs of fear which included hard breathing, fast heart rate, and the increase in blood acidity. For as long as a month after all training had stopped the monkeys continued to show high blood acid levels at the exact time in the day they had been subjected to unpleasant experience. It appeared that the body remembered the time of fear and continued to anticipate and react at that same time each 24-hour period even though the frightening situation had been removed.

When the acid-base equilibrium of the body is thrown out of balance, it usually is a sign of illness. This condition sometimes occurs in diabetes and kidney or liver disease. Diabetics or other people suffering from low blood sugar often suffer depression along with acid imbalance. A person with blood acidosis also metabolizes adrenaline faster than a healthy person.

This chain of circumstantial evidence suggests that some diabetic patients, for example, may suffer symptoms which do not exist around the clock but rather occur at certain times of day as a response to an unpleasant fear conditioning at some time in the past. Diagnosis at this level of refinement would require medical measurements taken around the clock for a number of days. In this manner, base line measurements of circadian rhythms may become an essential part of computerized medicine of the future.

One question requiring more study is whether conditioned responses to unpleasant emotional situations continue to resonate like periodic echoes of a bell, or ripples on a lake, becoming fainter and fainter until they die out, or whether the rhythmic response to old emotional conditioning tends to reinforce itself and intensify a growing neurotic condition.

Transient symptoms such as an illogical anxiety which occurs only at certain hours of the day could account for some of the apparent hypochondriacs whose ailments vanish mysteriously by the time they reach the doctor's office.

In a chain of investigation extending over four years, Dr. Stroebel and his associates slowly unveiled the interesting relationship between biological time and learning. They found, in essence, that the more intellectual or cerebral the task, the less difference it makes when it is learned. It is "gut" learning, experiences most loaded with emotion, that appears to be locked in with biological rhythms. Unavoidable anxiety is most influenced by the time of day. On the other hand, time of learning is less important in situations which can be circumvented. A purely cerebral discrimination situation, without pleasant or unpleasant connotations, showed no relationship at all with internal cycles.

Learning unavoidable fear relates to the time of day when adrenal steroid concentrations are highest in the blood. In man the maximum time of susceptibility for acquiring a neurosis-fear (and perhaps for unlearning it as well) appears to be the early morning.

If a person is repeatedly exposed to anxiety at 8 A.M. each day, and later tested at random times, he will show strongest fear at 8 A.M. An example might be a man who dreads to encounter the unreasonable demands of a boss every day at a morning staff meeting. Over a period of time his repeated exposure to unavoidable anxiety at his most vulnerable biological time of day could lead to neurosis or psychosis. Even if he quit his job to escape the situation, such a man might suffer internal and apparently illogical symptoms at the same hour of day for months or years after. In order to extinguish the anxiety neurosis, it may prove beneficial for a psychiatrist to measure his patient's adrenal rhythm and utilize the circadian factor in helping the man unlearn his fear.

A number of researchers consider adrenal rhythms as a sort of internal alarm clock which triggers emotional responses according to times when they were learned. It may be that certain people use these internal clues to "set" their minds to awaken at a certain time in the morning or to take a nap of a few minutes duration. If this is so, it might someday be possible to train people to "listen" to the rhythms of their internal clocks and learn to control their behavior.

. . . How Stress Upsets the Biological Clocks

WHILE consideration of biological time is important to learning and unlearning in emotional situations, stress may develop into neurosis or psychosis by upsetting the rhythms themselves. This was demonstrated through another study by Dr. Stroebel's group at Hartford in which monkeys were placed in a situation so frustrating it literally drove them mad.

The monkeys spent six hours each day in a "problem" cage equipped with two levers, one on each side of a hopper into which food pellets could drop. The animals used the righthand lever to press for rewards as they worked out a variety of problems. The lefthand lever, however, was a special one. During the weeks of tests, the animals were subjected at random times to uncomfortably high temperature, brilliant lights, raucous noise, or electric shock. If a monkey (accidentally at first) pressed the lefthand lever, he could rid himself of his annoying problems. As the weeks passed, the monkeys learned to treat the lefthand lever as a symbol of security. Some held onto it, refusing to let go, as a "security blanket."

After several weeks the lefthand lever was retracted into the wall. The noxious distractions were removed at the same time, but the monkeys frantically worked for hours trying to get at their "comfort" levers. In failing they became mentally disturbed. Some developed high blood pressure, asthmatic breathing, stomach upset, blood in the stools, and skin eruptions. Others became apathetic and performed tasks unpredictably. They stopped grooming themselves and their fur became mottled and matted. For hours they would pull their hair, masturbate, or catch imaginary insects. In short, the monkeys became either psychosomatically ill or psychotic.

Those monkeys with psychosomatic illness also tended to become desynchronized from their normal rhythms. Their brain temperature cycle remained roughly circadian, i.e., twenty-four hours long in period, but shifted out of phase with their environment. The animals with psychotic symptoms showed a different cycle change. Brain temperatures shifted slowly to a 30-hour cycle and then jumped abruptly to more than forty-eight hours, although behavior did not follow the same period.

Some investigators speculate that psychosomatic illness and some periodic diseases result when some of a person's internal clocks begin to run freely, moving out of phase with each other and with normal hours of sleep and action. Emotional stress, such as that demonstrated by Dr. Stroebel, may produce such an uncoupling of rhythms.

.. Using Electronic Instruments to Measure Human Rhythms

ANIMAL studies help to map the way to understanding human behavior. By means of the delicate electronic instruments developed to measure animal responses, it may become possible to measure human rhythms without confining people to the hospital or laboratory. Temperature cycles, the key to many others, could be measured by tiny thermistors planted under the skin. Then, by means of a small radio transmitter, the hour-by-hour readings could be transmitted to a central location and analyzed by computer while the person being studied goes about his normal business. That thought suggests the possibility that someday a person might be so monitored on a permanent basis with a telemetered feedback radio system to warn him when his rhythms are slipping out of phase and enabling him to make adjustments.

That day of rhythmic diagnosis of mental problems before they become severe may be far in the future, but current studies are charting new territory.

"The questions about circadian responses in human beings—to learning, to drugs, to the extinction process of psycho-therapy—still fall into the shadow of the great unknown," Dr. Stroebel commented. "Is the human being infinitely adaptable to any sort of schedule, to any variety of irregular habits—or is he a creature of structured time like his cousin mammals, the apes?

"The role of rhythms in man's well-being has not been demonstrated, but like the suspicion that a lack of rhythmicity spells illness, animal data make one suspect that well-being in man also depends upon a knowable harmony of internal rhythms coupled with a rhythm of overall behavior, synchro-nized in turn with the alternation of day and night—roughly every 24 hours.

"Such a harmony would have had a long evolution," he added. "It is in this shadowy realm, this elusive realm of invisible timing, that many of the interesting questions of psychiatry and psychology may be answered." [8]

> Sophisticated work by experimental psy-
> chologists indicates that emotion, learn-
> ing, and memory may vary in strength and
> weakness according to the specific time

in the biological cycle when they are experienced.

This relationship to human rhythms may be used to chart best individual times for learning, assist a person in understanding and coping with his own emotions and fixations, and offer a new realm for diagnosis and treatment of mental and psychosomatic illness.

CHAPTER 7

*Theories of disease
and aging. Biological rhythms
as aids in diagnosis,
desynchronized clocks as causes
of illness and death.*

What Happens When the Clocks Run Askew

IN 1796 MARY ANN LAMB, sister of the great English essayist Charles Lamb, stabbed her ailing mother to death in a fit of temporary insanity.

Mary was of high intellect and had assisted her brother in some of his writing. Both loved their mother with whom they were living at the time of the matricide—a tragedy which is possibly the most notorious in all the records of illnesses which occur with unexplained periodicity. Beginning at age thirty,

Mary had suffered psychotic attacks of violence and rage which recurred without explanation at predictable times.

Knowledge that her attacks of insanity came on in such fashion saved her from criminal prosecution. A lawyer friend was able to have her assigned to her brother's custody. Thereafter, at the first sign of irritability, her brother would rush her to the hospital or put her in a straitjacket. After recovering each time Miss Lamb was completely normal, going her usual rounds of entertaining literary friends and writing books or stories until the next attack.

Mary Lamb's periodic insanity persisted for fifty years. During that time she suffered thirty-eight attacks, but lived a normal, fruitful life in the times between. She lived with her brother until his death in 1834. She died thirteen years later.[1]

Old records of medicine and psychiatry contain many case histories of cyclic maladies, both physical and mental, which mystified scientists of their day and might have been more easily understood in light of today's knowledge of biological clocks. Periodicity and lunar influence in human illness was studied by the British physician Richard Mead in the eighteenth century. In 1704 he wrote:

"I know a gentleman of tender Frame and Body, who having once, by over-reaching, strained the Parts about the Breast; fell thereupon into a spitting of Blood, which for a year and a half constantly returned every New Moon, decreasing gradually, continued always four or five days . . ."

Mead also noted that people suffered epileptic fits according to phases of the moon. "The Girl, who was of lusty full Habit of body," he wrote of one patient, "continued well for a few days, but was at Full Moon again seized with a most violent fit, after which the Disease kept its Periods constant and regular with the tides; She lay always Speechless during the whole Time of Flood, and recovered upon the Ebb. . . ."[2]

Dr. Curt P. Richter of Johns Hopkins University, veteran investigator in the field, tells of an English soccer player whose knee joints became swollen and painful for two or three days at periodic intervals. Physicians tried to identify the cause of the recurring disability without success. They finally determined that the young man's attacks developed every nine days without apparent cause. Between times he seemed in top athletic condition. This knowledge did not help cure the affliction, but once its cyclic nature was defined and predictable, coaches were able to schedule games far in advance for the dates on which he would be able to play.

Another man suffered attacks of peptic ulcer every 140 days for more than 10 years. Once he had charted the long but predictable cycle, he was able to anticipate the ulcer attacks, watching his diet and reducing tensions to ease an attack he knew was certain to come.[3]

The cases of Mary Lamb, the patients of Dr. Mead, the soccer player, and the ulcer patient are clear examples of disability and illness which attack with periodic regularity in some long cyclic coincidence of synchronization with the internal rhythms of body and nervous system. Such cases, along with advancing research in biochronology, have focused the attention of investigators upon three areas in which internal clocks may be involved in causing, or assist in diagnosing, ailments and disabilities.

The first includes those illnesses which recur periodically through no otherwise explainable cause. Second are disabilities or sickness which result from throwing the rhythms out of phase. Third is the growing ability to "read" abnormal rhythms in diagnosing disease or to predict its onset. In man, more than 100 functions and structural elements oscillate between maximum and minimum values once a day, ranging from temperature to mood and mental performance. The interlocking and synchronization of these circadian oscillations, along with shorter or longer cycles, may well determine whether a man is in balance with time and how well or badly he is able to function when the balance is disturbed.[4]

. . . Cyclic Maladies

THE moon often was blamed for periodic maladies in ancient times. One such was the Italian *chiodo lunare,* or moonstroke, a neuralgic pain around the eye socket believed to occur when the moon rose. It would disappear as the moon set. From Richard Mead in 1704 we learn that:

> Epileptical diseases, besides the other Difficulties with which they are attended, have this also surprising, that in some the Fits do constantly return every New and Full moon; the Moon, says Galen, governs the Periods of Epileptick Cases . . ." [5]

Also from Galen, of a time shortly after Christ lived, we have circumstantial evidence of moonstroke and moon madness.

In the Middle Ages health, strength, and sexual power were supposed to vary with the waxing and waning of the moon. Moonlight was believed to cause lunacy, confer beauty, or cure warts and some diseases. Menstruation was connected with the lunar cycles. To let blood when the moon and tides were full was considered a bad practice and, as in the case of Shakespeare's Falstaff, it was common superstition that death occurs most often at the turning of the tide.[6] (Even Christiaan Barnard, the South African physician who performed the first human heart transplant in 1967, noted that "most people die at dawn.")

Nowhere is the lunar cycle more obvious in peak and decline of health than in the female menstrual cycle. Although periodic depression and elation are accepted by most women as their curse in life, the premenstrual syndrome is the most common of all cyclic discomforts. Premenstrual tension is the catch-all phrase for many symptoms which occur generally in the four or five days preceding the onset of menstruation.

About 60 percent of all women suffer noticeable change at this time. In some it may be no more than mild irritation, depression, headache, and decline in attention or vision. Others become jittery, weep, sleep poorly, suffer dizzy spells and even temporary nymphomania. In most women, the malaise is considered to be a minor maladjustment of the normal cycle of endocrine gland secretions (which also may account for some other periodic illnesses). Whether the disorganized rhythm is slight or severe, it weakens a woman's resistance and intensifies other illnesses. In severe cases these can range from respiratory ailments to the activation of chronic disease such as arthritis or ulcers. Many women complain of stomach trouble and altered appetite, and show sharp changes in blood sugar levels.

Monthly changes in water retention may account for premenstrual headaches and blurred vision. The high proportion of virus and bacterial infections which strike at this time may be related to internal cycles of the hormones estrogen and progesterone, which are important for fighting infection.

The premenstrual syndrome, however, becomes a more serious personal and social factor in the psychological and behavioral changes it brings about. At this time of month women are most likely to be admitted to psychiatric wards. One prison study indicates that violent crimes by women tend to cluster within their special days of stress. This is borne out by other surveys which show that 63 percent of crime by women in England, and 84 percent in France, occurs in the premenstrual period, along with a disproportionate number of suicides and accidents and a decline in intelligence test scores, visual sharpness, and quality of schoolwork.

In the United States, the loss of professional work, when women are absent on sick leave from their jobs during the "curse," is valued at about $5 billion per year.[7] Some sociologists suggest (while dodging the brickbats of women's liberationists) that the primary reason women have not achieved full equal status with men stems directly from the premenstrual syndrome, which takes them away from professional jobs with predictable regularity. Since their moods are unpredictable at such times, even if women remain at work, the business world fears female judgment on important decisions and this may hamper the climb of most women to executive positions. It is this same periodic weakness in women which causes men to treat them as the "weaker" sex and elicits whatever shreds of male gallantry are left in protecting their women against the sharper edges of the professional world.

It is strange that a natural cycle should impose such complexities upon the workings of society, but women of enlightened America should know at least that they are not alone. Dr. Oscar Janiger at the University of California at Irvine made a pilot study of Lebanese, Apache, Japanese, Nigerian, Greek, and American girls. He found that premenstrual tension seems to be universal among women of diverse cultures. To obtain wider and more precise data, Dr. Janiger extended his investigation to the higher primates. From questionnaires filled out by zookeepers he learned that female rhesus monkeys, chimpanzees, and gorillas also show premenstrual symptoms of lethargy, irritability, and belligerence.[8] This was not an effort to compare women with monkeys but to identify suitable experimental animals through which researchers someday may find ways to alter the internal rhythms and ease the age-old curse of womankind.

The internal rhythms of man are caught up in the mysterious web of time synchronization in many other diseases including allergies, kidney disease, epilepsy, tuberculosis, diabetes, cardiac illness, and glandular diseases.

One baffling ailment is a form of high blood pressure which recurs regularly. This was first brought to medical attention in 1953 by a tiny, shy nun who suffered cyclic bouts of fever coupled with intense headaches. These at first lasted only a few minutes but later recurred and lasted for twelve hours or more. Her symptoms were variously diagnosed as malaria, migraine, anxiety neurosis, and trichinosis—none of them correct. Finally it was observed that she suffered hypertension (high blood pressure without detectable cause) only during her spells of fever. Mathematical analysis by Dr. Halberg of Minnesota revealed that the nun's episodes of illness repeated themselves every eleven days.

She was studied more extensively at the National Institutes of Health, where it was learned that she underwent wide changes in concentrations of adrenal hormones, one of which regulates retention of sodium in the system. This case has spurred doctors and scientists to search for the internal clock or clocks which regulate the sudden increase and decrease of blood pressure in hypertension patients. Measurement of an individual's circadian rhythms is likely to become an essential part of medical diagnosis for this form of high blood pressure.

Attention to standard rhythms and their disturbance may be equally important in other medical cases. Dr. Werner Menzel of Hamburg, Germany found in a child with Hodgkin's disease (a malady of the lymph system) that body temperature followed a 12-hour instead of the normal 24-hour rhythm. He also discovered 12-hourly peaks in urine volume in patients suffering from arterial tension and oversleeping. People with liver disease often showed peak temperature and urine excretions at night instead of by day.[9]

... Using Rhythm in Diagnosis and Treatment

DR. J. N. Mills of England has compiled many instances in which abnormalities of circadian rhythm may symptomatize or contribute to the cause of certain ailments. A five-degree temperature swing was found in a patient who suffered a food allergy. Abnormal rhythm of vital capacity is common in people with tuberculosis. A large rhythmic variation in blood pressure in patients with acute glomerulonephritis (a kidney infection) seems to subside during recovery. Dr. Mills points out that the common occurrence at night of cerebral hemorrhage, pulmonary edema (swelling with accumulation of fluid), and cardiac asthma is due to changes that make up part of a natural circadian rhythm, and the same probably accounts for the unequal overall incidence of death, which is highest early in the morning.[10] Such observations add little by little to the growing new set of analytical tools available to physicians for determining subtle variations behind grosser symptoms of illness.

More than 2000 years ago Hippocrates, the grandfather of physicians, noted:

> So in one place the blood stops, in another it passes sluggishly, in another more quickly. The progress of the blood through the body proving irregular, all kinds of irregularities occur.[11]

The great healer had few experimental facts to back up his observation, but recently a team of clinicians in Florence, Italy detected a circadian rhythm in the blood flow in lower legs of people with peripheral arterial disease. Normally the circulation of blood seems to be independent of any clock except the heartbeat; so this periodicity may help explain some of the recurrent symptoms of this form of artery malfunction.

A Swedish doctor noted some years ago that patients with ulcers or cancer of the stomach complained of hunger pains, which could be suppressed by eating a meal. Some patients, however, were irregular in their complaints. In an extended study of 200 people (83 had cancer, the remainder ulcers) the doctor found that almost all of the cancer patients showed irregular rhythms of sodium chloride excretion which seemed to coincide with the irregular hunger complaints. None of the benign ulcer patients showed this irregularity. The point of such findings is that a clinician who "listens" to the biological clock by observing patient complaints and noting the hour may be more accurate in detecting the difference between cancer and stomach ulcers than a doctor who resorts to X-ray or exploratory surgery.

At the University of Florence researchers in the biological time structure of illness (called *chronopathology*) attempted to determine the difference in internal rhythms between fifteen healthy people and eight with peptic ulcers. Doctors earlier had conjectured that stress, expressed through adrenal hormones, might be a critical factor causing stomach ulcers, but early tests failed to find a correlation. The Italian study, which followed the body rhythms through a full 24-hour cycle, detected that a difference in cortisol excretions between people who have peptic ulcers and those who do not indeed exists.

It has been shown previously that the suffering of ulcer patients may be linked with the 90-minute cycles of sleep. Normally the level of adrenal corticosteroids in the blood seems to rise toward morning in cyclic waves corresponding to the brief periods of REM or dream sleep. Since adrenal hormones are secreted in such a pattern, they may be the agent responsible for triggering the excessive amounts of gastric acids which intensify ulcer suffering.

Partial verification of the hypothesis came from the University of California at Los Angeles, where investigators recorded nightlong brain-wave patterns of healthy people and duodenal ulcer patients, all of whom slept with stomach tubes inserted to measure the output of gastric acid. In the ulcer patients the times of secretion corresponded closely with the brief periods of dream sleep. The healthy subjects showed no periodic secretion whatever. Thus evidence shows more clearly that an evolutionary device designed to prepare early mammals for response to danger has become a hindrance in treating the modern affliction of stomach ulcers. The corollary suggestion follows that if the dream sleep cycles and their related oscillations of adrenal hormones could be suppressed by drugs or other therapy, it might help to hasten the recovery of some ulcer patients.[12]

A more complicated interrelationship of biological clocks is involved in the onset and progress of sugar diabetes (diabetes mellitus). This disease afflicts millions of people who suffer high levels of blood sugar with excessive thirst, abnormal weight changes, lack of energy, and faintness. If not controlled by therapy or diet or both, the disease can lead to severe acidosis, blindness, coma, and death. It is well known that diabetics lack insulin (a secretion of the pancreas), which is needed to convert sugar into useful forms. Until recently doctors routinely prescribed regular doses of insulin to control the disease on the assumption that the body needs the same amount of the hormone at all times. Now it appears that more complex cyclic factors contribute to the ailment.

It begins with glycogen, a form of animal starch stored in the liver and there transformed to the type of sugar used by all cells of the body. If liver glycogen does not reach normal levels at the right time, as in some liver diseases which disturb the cycles of production, the entire system—most critically the brain—is deprived of energy during times of daily demand. Glycogen levels usually are related to food intake, but scientists have found that a circadian rhythm persists even in a person who is fasting, which offers a clue to the nature and treatment of diabetes. Dr. Arne Sollberger of Yale University discovered that the glycogen curve descends in late afternoon and continues falling during the night so that by morning the liver has used up most of its glycogen.

In a healthy person the carbohydrates and proteins of food are transformed with phosphate to form adenosine triphosphate (ATP), the energy-bearing unit within each cell. This process is aided by the pancreatic hormones, insulin and glucagon, and a hormone from the pituitary gland. The diabetic may suffer hunger when he is well-fed because he does not produce enough insulin at the proper time in the 24-hour cycle to break down the sugars he eats into usable forms such as glucose. The brain requires constant nourishment in glucose from the blood. Since the diabetic is not producing it correctly, he may suffer brain starvation in the morning or after other fasting because the hormonal balance or cycle has been disrupted. When liver glycogen is low, a diabetic sufferer without food might respond badly to a dose of insulin.

Investigators in Paris found there is a circadian rhythm (among healthy patients) in blood insulin and blood glucose. This suggests that refinement of diabetic therapy would include administration of artificial insulin not on a constant basis but according to the cycles of other biological clocks that are out of balance.

These include the rhythms of different cells produced by the pancreas. One forms digestive enzymes which break down fats and sugars into useful components. Another produces glucagon, which increases available blood sugar by transforming glycogen into glucose. The third type of cell produces insulin which, in turn, helps to regulate glycogen and lowers blood sugar levels. The circadian rhythm of mitosis (division) of these three types of cells operates on different schedules which must be synchronized. Thus proper treatment of diabetics should aim at restoring the correct interrelationship of these rhythms to reproduce the proper rise and fall of insulin levels, rather than merely increasing deficient hormone levels.

Other studies reveal that lag times between overlapping of cyclic peaks of adrenal hormones, fatty acids, and the body's rate of glucose disposal may be clues to how sugar diabetes begins. The current picture of the disease is that it involves the biological clocks governing several glands and may be triggered by mistiming in a region of the brain by inherited predisposition.[13]

The complexities of synchronizing biological time in relation to diabetes exemplify how the measurement and analysis of circadian rhythms may open new vistas in diagnosis and treatment of many endocrine and infectious diseases.

More startling is the recently discovered role of biological clocks in the parasites which cause several debilitating diseases in man and other mammals, especially in the tropics. One of these is filariasis, more commonly known as elephantiasis because of the gross deformities it produces in lymph glands and scrotum or legs and arms. This disease is caused by nematodes, or tiny worms, which are transmitted in larval form to the bloodstream by insect bite.

One species of parasite worm, or microfilaria, congregates more numerously in blood near the surface of the body during the night, while accumulating in lung capillaries during the day. Another species follows an opposite circadian rhythm. In both cases, the cyclic behavior is an evolutionary timing mechanism developed to insure reproduction and survival. Each of the microfilariae maneuvers so as to be in the blood near the skin's surface at the right time of day or night to meet the vector insect which carries it from one victim to another.

At first, investigators believed the parasites simply responded to the circadian rhythm of the victim in which they were proliferating. It was found, however, by changing and reversing the schedules of infected people that the microfilariae quickly adapted to the changes. This proved the parasites have their own precise biological clocks acting according to certain triggers or zeitgebers (such as temperature) within the host. Such knowledge suggests to physicians that small changes in oxygen or carbon dioxide tension or adrenal hormones, applied regularly for many days, may be more effective than a single powerful stimulus in helping the body rid itself of the disabling parasites.[14]

Once it is understood that a rhythmic sense of timing is essential to almost every function of the brain and body—from emotions to the division of individual cells—then it becomes logical to ask what role the biological clocks may play in cancer, the most dreaded killer of all.

Cyclic order and disorder indeed are involved in cancer. The most mysterious finding in this regard goes back to Dr. Janet Harker at Cambridge and her painstaking experiments to find the seat of the biological timer in the cockroach.

Dr. Harker first tracked the source of the insect's rhythmic behavior to the subesophageal ganglion, an organ the size of a pinhead forming the secondary of the insect's two brains. Then she used a microscopic cauterizer to shave away the ganglion, a few cells at a time, until finally she found just four cells which were responsible for maintaining the cockroach rhythm. The cockroach is the only creature in which the biological clock mechanism has been isolated with such precision.

The patient experimenter then sought to learn how the cockroach's timer worked. One possible method was to throw it out of kilter. With pioneering microscopic surgery, she planted in a normal cockroach the ganglion (with clock) which she had removed from another cockroach that had been trained to function opposite to its normal circadian rhythm. The receiving insect thus was exposed to two biological clocks, each giving diametrically opposed directions. These experiments were repeated many times with many insects.

The result was totally unexpected. Under the stress of exposure to two clocks running out of phase with each other, the cockroaches developed intestinal cancer and died.[15]

"The discovery of the location and transplantability of a biological clock supplies us with an invaluable tool for further research," said J. N. Berrill, then professor of zoology at McGill University. "Most surprising is the discovery that transplanting a neurosecretory clock into another cockroach when they are out of phase causes a malignant tumor to grow beneath the donor clock. Thus the search for the clock in the cockroach has led to more mystery than we started with. We now have an unexplained association of a clock with cancer." [16]

It is tempting to conclude that if the stress of violent disrhythm can give cockroaches cancer, then disturbing human rhythms by jet travel or otherwise might be a contributor to cancer in us as well. However, that is too great a jump to make from one piece of experimental evidence in an insect to the mass of unknowns involved in identifying and manipulating human rhythms. In the meantime, there is more nearly related evidence of the role of biological rhythms in human cancer.

Most apparent, though not fully understood, is the violent "unnatural" speed with which cancer cells divide and multiply. This scourge of humanity, and many related species, is one of the great puzzles of medical science. It is more accurate to consider it a multiplicity of puzzles because many forms of malignancy attack many parts of the body. Many substances have been found which can trigger the onset of tumors in laboratory animals. But research has not determined how and why cells in living tissue follow well-disciplined rhythms in division and rebuilding the body, then one day change shape and form and begin spreading and growing in wild, uncontrolled and destructive fashion. Some cancers suggest that an alien beast has taken control of our tissue and then grows itself within us.

Where biological clocks are concerned, investigators have discovered that cell division in cancerous tissue proceeds at an abnormal rate seemingly unregulated and out of phase with the circadian period of mitosis rhythm in normal surrounding tissue.

Dr. Halberg at Minnesota has determined that normal cells divide most rapidly during certain intervals of the 24-hour day. In humans, most rapid division occurs at night when we are sleeping. Scientists also found the rhythm could be altered if light-dark periods were shifted. This suggests that a rapid and drastic change in living schedule might affect the way human cells produce protein and reproduce themselves in the lifelong process which keeps our internal organs functioning. Dr. Harker's cockroaches suggest that a sharp stress placed upon the interlocked biological rhythms may lead to the violent cellular disrhythm of cancer.

In one test of human cancer tissue, Drs. Halberg and C. P. Barnum collaborated with Dr. Mauricio Garcia-Sainz at the Oncological (cancer study) Hospital in Mexico City. One group of tissue samples was taken at two-hour intervals around the clock from cancer patients before they went into X-ray treatment. These were analyzed for rhythmicity of cell division and compared with a second set of samples collected after the patients had been treated.

By direct count of dividing cells and data analysis, the investigators determined that the cancer cells before treatment were not circadian in periodicity. Cell division rose and fell in a 20-hour rhythm in some cases, an 8-hour cycle in others. Surrounding healthy tissue in all cases followed a normal 24-hour cycle of cell division. After X-ray treatment, the rhythm in the cancerous tissue changed toward normal. According to Dr. Halberg, the lack of normal cycles in cancer cells suggests that a defect in time integration creates a form of timing anarchy leading to tissue abnormality.

Other tests were devised to find what timing changes, if any, occurred by inducing cancers in laboratory mice either by using a cancer-triggering compound or implanting cancerous tissue. The investigators found in the laboratory animals (destined to develop tumors of the mammary glands) that even before the growth started there were changes in the rhythm of cell division in tissue taken from the ears. Traces of abnormal cell behavior, long before the grosser signs of a tumor appear, offer the possibility that human cancer might also reveal itself in its earliest stages through testing the circadian rhythm of cells at the surface of the body. At today's practicing

level of medicine and surgery, early detection remains the
strongest weapon against cancer.

In his work with Dr. Garcia-Sainz in Mexico, Dr. Halberg has
attempted to learn the characteristic abnormal rhythms of
different forms of cancer. By clocking the cycles of malignant
cells, researchers hope it may become possible to time X-ray
therapy to strike at specific cancers during their most vulnerable phase while avoiding or diminishing the side effects of
cell destruction in nearby healthy tissue.[17]

. . . Periodicity in
Mental Illness

THIS chapter contains only representative examples of the
importance which irregularity of biological rhythms implies in
physical illnesses, their diagnosis and treatment. The cycles of
life become even more apparent in the study of mental illness.
Some forms of mental disturbance show periodic swings from
normal to abnormal every forty-eight hours. Others exhibit
cycles spanning weeks or months. The tragic case of Mary
Lamb is one of the latter.

Some of the most careful studies of periodicity in mental
disability have been conducted in Oslo, Norway by Dr. Leiv
Gjessing and his father, Rolv Gjessing, at the Dikemark Hospital. There, since 1905, patients have been kept under care in
pleasant surroundings and not isolated from the surrounding
community. In this atmosphere the doctors have conducted
long-term studies of people in their predictable swings from
normalcy to excitability, violence, and finally catatonia, the
mute and frozen state of mental illness which resembles
paralysis.

As early as the 1920s, Dr. Rolv Gjessing observed some patients suffered stupors every two weeks but behaved normally

at other times. When their behavior changed, so did their physical appearance and some internal functions as well. During catatonic stupor one man's saliva was so thick it could be drawn out like chewing gum, and his skin became oily. Another's body would turn rigid and he would babble maniacally before going into the trance. High pulse rate and blood pressure indicated that, though withdrawn from the external world, internally he was undergoing great activity and stress. Recollections of one patient described hallucinations such as those encountered in the use of mescaline or LSD.

The doctors, from their father-to-son observations, believe the cyclic nature of these attacks indicates that flaws in metabolic rhythms may be the underlying basis for mental disease. Periodic psychosis often begins in the early twenties and sometimes abruptly. Dr. Leiv Gjessing speculates that stress, brain damage, or perhaps a metabolic shock might damage one of the biological regulators, thus producing the cyclic symptoms.[18]

Schizophrenia (a type of mental illness in which the patient loses contact with the world around him) and manic-depressive psychosis (a form of mental swing from mania to depression) both occur in cyclic phases. People who suffer these disabilities usually are not sick all of the time but have attacks at intervals, sometimes totaling no more than two months out of a year. The eating habits of such people change drastically from normal to abnormal periods, and some doctors feel that the two forms of illness could be triggered by imbalance of the internal cycles governing metabolism.

Dr. William Bunney at the National Institute of Mental Health has studied the "switching" point in manic-depressive patients, finding that the change from depression to mania is preceded by increases of norepinephrine (an adrenal secretion) in the brain. During the twenty-four hours before the change, the patients also showed a decrease in normal periods of dream sleep.

Some manic-depressives change condition on a 48-hour cycle. One such is a former English boxer studied by Dr. F. A. Jenner of Middlewood Hospital in Sheffield, England. The fighter began suffering his mental illness after an accident. For twenty-four hours he is overactive, talkative, and sometimes irritable with grandiose ideas about science and the world. Then, sometime in the night, his mood changes and he awakens lethargic and bleak and has trouble getting up. On his inactive days he urinates and excretes more, but eats and drinks less than on his days of mania. Studies of this man over eleven years have led doctors to believe his cyclic disease shifts its pattern because of dissociation of chemical excretion rhythms.

Dr. Curt Richter at Johns Hopkins has hypothesized that shocks, such as infection, allergy, surgery, or physical and emotional stress, may upset the overall coordination of phase relations among the many body clocks, setting up metabolic rates out of synchronization with others and thus accounting for the periodic "beat" or oscillation to the symptoms of illness.[19]

. . . The Rhythm of Dying

AFTER considering some of the ways in which our internal clocks regulate health and illness, is it possible to consider death as the ultimate disease? Is our total life-span the longest natural cycle of all, differing in each individual according to hereditary and environmental imprint? Is there a master clock which runs down at the hour of our death?

Answering such a question would require detailed study of many people from birth to death with a record of every traumatic experience, injury, or illness suffered during their lifetimes. Most research programs are too short to complete such a study, but someday it may be done.

Most people tend to think that after maturity there is a long plateau of life stability and then a rapid, accelerating downhill slide into old age. According to Dr. Nathan Shock of the National Institutes of Health in Baltimore, however, "for most functions no adult plateau exists. Most changes seem to be gradual and progressive. In terms of a theory of aging, the body dies a little every day."

We know there is a distinct circadian rhythm to the pace at which cells divide in their rebuilding of our body tissues. Is it this rhythm which slows down, or is altered, when we grow old?

In the view of Hans Selye, each person at birth is endowed with a finite storehouse of "adaptive energy" which he may use through his lifetime to cope with various forms of change and stress. If a man's store of adaptive energy is small and if he is forced by life into a position where he must work hard and stressfully, he probably will lead a short, unhappy life. On the other hand, with a great deal of adaptive energy, and the opportunity of working under less stressful conditions, life should be comparatively long and happy. In this view, adaptive energy is like a bank account that you use up by making withdrawals, but that you can never increase by making deposits. How you use it in the framework of life's natural rhythms has much to do with how well and how long you live.[20]

What causes cells to stop their rhythmic job of reproducing themselves still is open to conjecture, but according to Dr. Charles H. Barrows, biochemist and gerontologist at Johns Hopkins University, there are four general hypotheses as to the cause of aging. Oldest, and least substantiated, is the theory that each cell is endowed at conception with some fixed amount of vital substance that is used up over time. When the substance is depleted, the cell dies. Researchers never have found such a "vital substance," but there is relationship here with the adaptive energy concept.

A second theory suggests that damaging substances accumulate in the cell and interfere with its normal function.

Third is the so-called error theory focusing on the genetic material within the cell, DNA and RNA, which carry all the hereditary information that goes to make up a specific individual. By this theory, alterations or errors accumulate in the cell's DNA molecule with passage of time and from various causes. Such errors, it is postulated, cause the cell to produce defective enzymes which in turn lead to death of the cell.

The fourth theory—most widely accepted at the present time—is that man's total life-span from conception to death is programed in the genes which carry the hereditary message. This theory says, in effect, that the fertilized egg has stamped into it a timetable not only for embryonic development, birth, the appearance of baby teeth and a child's first step, but also the cycle of reduced strength, wrinkled skin, susceptibility to illness and death—the final disease of aging. If this theory is true, then the process of aging is truly no more than the downswing of a long cycle which began its upward climb at conception in the womb.[21] Although this rhythm might be subject to change, phase shift, or alteration—as are most internal clocks—the hypothesis suggests that the rhythm of life toward death may be preset in each individual and perhaps synchronized with all other cycles from heartbeat and brain waves to the multitude of secretions which follow a 24-hour day.

Other studies suggest that aging and death, rather than the culmination of a long cycle predetermined by evolution, is the result of other body clocks gradually slipping out of synchronization. The idea of dissociated rhythms as a contributing cause to ultimate decay is easier for scientists to live with than the notion that life might be a long predestined cycle set internally by hereditary clocks or externally by shifting forces of the universe.

Italian researchers in Florence have suggested that the decline of sex hormones could influence the overall timing of a person as he grows older.[22] This would dovetail with the concept that evolution cares nothing about individuals in a species once they have passed reproductive years and indeed, may program death after that stage to prevent overpopulation.

Such hypotheses defy scientific methods at clarification, but there are indications that understanding circadian periodicity may become an essential factor in geriatric medicine and help explain the sensitivity of older people to slight changes in schedule, alterations in light, temperature, or noise, and drug dosage.

Attacks of many illnesses, both mental and physical, occur with periodic regularity, apparently linked to the malfunction or desynchronization of biological cycles within mind and body.

Other illnesses may result from throwing internal rhythms out of phase. Knowledge of normal and abnormal rhythms promises a new set of internal timing measurements which doctors may use in diagnosing the early onset of disease and treating it more effectively. Some beneficial findings include:

1. Hormone cycles may offer a clue to the cause of stomach ulcers, and sufferers might be eased or cured by dampening the dream cycle during sleep.

2. Diabetes involves biological clocks governing the cyclic excretions of several endocrine glands. Correcting cycles that are out of synchronization may become

151 . . .

more effective than regular doses of insulin or drugs in treating the disease.

3. Parasitic diseases, such as filariasis, may be treated by altering the internal clocks of the parasites.

4. Cancer occurs when the rhythm of normal cell division goes unaccountably out of control. It may become possible to control cancer more effectively by attacking when its own abnormal cycle is at low ebb.

5. The periodic nature of many mental illnesses indicates they may result from damage or shock which throws internal glandular clocks out of synchronization.

6. Understanding the rhythms of cell division may illuminate the causes of aging and death.

CHAPTER 8

*The interaction of drugs
and body cycles.
New medical knowledge
to hit germs when they're down
and resistance is high.*

Rhythm-wise Medicine– Timing Drugs and Surgery

IT IS STANDARD PROCEDURE for the U.S. Food and Drug Administration to require a test of new drugs upon laboratory animals before they are approved for clinical test or human use. A lethal dose is determined by measuring the amount which causes 50 percent of the lab animals to die.

In one such study rats were kept on a steady day-night schedule and injected with potentially lethal amounts of the stimulant amphetamine at two-hour intervals. Around 6 A.M., when the animals were at the end of their activity cycle, only

6 percent died of the drug. When it was administered at midnight, peak of the rodents' activity, 77.6 percent died.[1]

This is an isolated example from a growing body of evidence showing that the biological time of day determines to a large degree the vulnerability of living creatures to disease, drugs, poisons, and other abnormal stress.

The time of day when a person becomes infected or takes a dose of certain drugs may make the difference—literally—between life and death.

This knowledge of oscillating susceptibility opens up several possibilities for human betterment, ranging from the control of insect pests with minimal environmental harm to a new strategy for treating illness. Understanding cycles of vulnerability may explain why some people suffer more savage attacks of a disease than others. It may help to control and cure addicts and alcoholics. It may determine how long an aspirin will kill pain or how well a barbiturate may help a person sleep. More importantly, it may help surgeons to schedule operations so as to enhance a patient's chance for life, or time radiation therapy and X-ray examination to take advantage of the hour when healthy cells are most resistant to damage.

"Although the existence of biological rhythms has been known for a long time, the number of different processes that undergo rhythmic change was not appreciated until recently," said Dr. Maurice B. Visscher of the University of Minnesota Medical School. "It might appear that this development would increase the complexity of bio-medical research. Actually, the contrary is true, because it is now possible to control a variable which was not previously recognized."

How the variable of proper timing may be made to work to our advantage may be demonstrated with as lowly a creature as the cotton boll weevil, one of many insects which destroy a large share of the world's food and fiber. Scientists at Texas A & M University dosed colonies of boll weevils with insecticide. One group had been conditioned to a 24-hour light-dark period with "dawn" at 6 A.M. Another group had its clocks shifted by artificial light and dark so that "dawn" came at 9 A.M.

When pesticide was applied to the first group at 6 A.M., it killed 20 percent, while only 10 percent of the second group died when exposed to the poison at its dawn, 9 A.M. In another test, when the insects were treated three hours before "dawn," 80 percent died. When the poison was administered three hours *after* "dawn," the mortality rose to 90 percent.[2]

The common housefly is most vulnerable to the poison pyrethrum about 4 P.M., while other insects are most sensitive to other poisons at different hours in their cycle. These findings by Dr. William Sullivan and other scientists at the U.S. Department of Agriculture begin to form a time-series pattern which may enable the nation's farmers to time the spraying of pesticides for maximum kill. This could result not only in more effective pest control, but also reduce the use of long-lived poisons such as DDT which some believe pose a threat to the ecological balance of the earth. As another example, knowledge of the mosquito's internal cycle of highest susceptibility could add a new dimension to the control of malaria in tropic areas of the world.

.. When Humans Are Most Receptive to Disease and Vaccination

EXTRAPOLATION of data provided by experimental animals indicates that if a person catches a cold at a particular time of day it might result in nothing more serious than a runny nose and minor discomfort. If contracted at another time in the biological cycle, it might develop into pneumonia.

Dr. Ralph D. Feigin at Washington University in St. Louis subjected mice to pneumonia germs at different hours in doses ranging from small to lethal. To his surprise, the physician found that the time of infection was as important to survival as the size of the dose. The time when mice showed their greatest ability to resist the infection was 4 A.M., when they were approaching the end of their activity cycle. Relating this to humans would indicate afternoon as our time of greatest resistance. In the early morning hours we may be most vulnerable to disease.

This could relate to Dr. Feigin's companion discovery of daily oscillation in production of amino acids. Such acids are the basis for creating proteins, necessary to all life processes. acid rhythm could be thrown out of phase with injections of a 8 P.M. in most people, but this would vary among those who do not follow the standard day-night cycle.

Dr. Feigin found that bacterial and virus diseases, as well as noninfectious illnesses, have dramatic effects on amino acid rhythms. This daily cycle also is upset by vaccination. Therefore he concluded that a person's biological time of day should be taken into consideration when determining the best time for inoculating against disease.

His opinion was based upon tests which showed the amino acid rhythm could be thrown out of phase with injections of a vaccine for a virus disease known as Venezuelan equine encephalomyelitis. Men immunized at 8 A.M. showed less disturbance of amino acid concentrations than did those who received the vaccine at 8 P.M.

"This work is so new that there is no way to make suggestions until we have completed further studies with individual vaccines, drugs and diseases," Dr. Feigin said. The work implies, however, that drugs as well as vaccines may be more effective at certain hours than at others. "The time of day a medication is given might well be a factor in why three out of five patients treated for the same disease in the same way survive, while the two others die," he added.

His work in the rhythms of vulnerability already has yielded knowledge of direct medical benefit in another way—an amino acid test for faster diagnosis of disease and especially for fevers of unknown origin.

High fevers are common among children and can have many causes. Before proper treatment can be prescribed, doctors must find out whether the fever is bacterial, viral, or non-infectious. In some cases this determination may require from eighteen hours to more than two weeks. That period may be shortened by the amino acid test. In one case a two-year-old child was admitted to St. Louis Children's Hospital with a 106-degree temperature. He had been treated with antibiotics without any apparent effect. Dr. Feigin's amino acid test quickly ruled out an infectious agent, enabling doctors to concentrate on noninfectious diseases with similar symptoms. They discovered the boy was suffering from inflammation of the colon. He was treated for this, recovered, and was soon discharged from the hospital.[3]

Basic research in the oscillating vulnerability to disease has been accumulating for almost thirty years. In the early 1950s Minnesota investigators dosed mice with bacteria which cause undulant fever. Injections proved least harmful during the rest period but were almost always lethal during the period of activity. An opposite 24-hour cycle was found in exposure to intestinal bacteria. Time maps of many diseases charted against cycles of vulnerability someday may become an auxiliary tool to be used with clock and calendar to plan a person's activities. If a child were known to be prone to colds and flu infection, for example, he could be kept away from other human contact during his periods of maximum susceptibility.

Dr. Halberg's work at Minnesota in this regard predated, but supports, Dr. Feigin's findings that the time of day is important for vaccination as well as disease exposure. This may be due to internal clocks which determine the cycle of the body's natural production of immunizing or disease-fighting agents— white blood cells and gamma globulin. Levels of gamma globulin reach a peak during the last six hours of daily activity and the low point at the end of rest.

Related tests with an anti-inflammatory drug revealed that when it was given at certain hours it caused severe loss in weight. This side effect of the drug, it was learned, could be eliminated by placing test subjects on a new light-dark schedule, similar to what happens when jet travelers fly several hours east or west. This cyclic weight reaction could be significant when it comes to prescribing medicine for infants or elderly people, in whom weight loss might be critical to survival.

. . . Cycles in Susceptibility to Noise and Epileptic Seizures

OTHER kinds of stress also may cause mild or severe results, depending upon the time of day they are experienced. One example is loud high-frequency noise, which can so excite an animal that it will go into convulsions and die of audiogenic seizure or spasm.

This reaction was tested under controlled circadian timing at the Minnesota biochronology laboratories. Mice were exposed to the ringing of electric bells for a minute at a time. During their daytime rest period, the mice responded by crouching or walking around as though they were offended. During their activity period, the same mice exposed to the same noise suffered audiogenic seizures and died. Their peak sensitivity corresponded to the top level in their daily body temperature.

Because of the similarity between audiogenic seizures and attacks of human epilepsy, doctors checked patients in a state hospital and found that their seizures clustered most frequently in the early morning hours. Brain-wave abnormalities were found at the time of day when the epileptic usually suffered an attack, even if none occurred that day. This suggests that the 24-hour distribution of abnormal brain activity is stable and predictable, which may help doctors understand the basic mechanisms of epilepsy and plan the most timely application of anticonvulsive drugs.

These studies also add another chapter of evidence to the growing dangers of environmental decay in industrialized society. Stresses from air and water pollution, and especially growing levels of noise, may intensify mental disturbances such as epilepsy, rendering it advisable to offer greater isolation and protection during an individual's most vulnerable time of day. Also, if an epileptic undergoes a drastic shift in his normal time schedule, such as jet travel or changing from a day to night job, this phase shift may increase his vulnerability to seizures at all hours of the day and night.[4]

.. Timing of X-rays, Drugs, and Other Therapy

NUCLEAR radiation is one of the primary tools employed in medicine to destroy and control the growth of malignancies, but radiologists always are inhibited by the fact that such radiation also destroys healthy tissue surrounding a cancer. Radiation therapy therefore always involves the question of whether more good than harm will result from the exposure.

The preceding chapter briefly discussed the work of Drs. Halberg and Garcia-Sainz in identifying the circadian periods when cancer cells are most vulnerable to attack while avoiding damaging side effects to healthy tissue. It appears now that healthy tissue also is more resistant to radiation at some times than at others.

In the 1960s Dr. Donald Pizzarello at the Bowman-Gray Medical School in Winston-Salem, North Carolina found that the dose of X-rays which made lab animals sick during the day would kill them during the night. Attempts to identify the mechanism behind this cyclic susceptibility were reported by Drs. Y. G. Grigoryev and N. G. Carnskaya of the Soviet Union. They concluded that deaths resulted principally from damage to the tissues in which blood cells are formed, bone marrow and spleen.

Translating these findings to man would indicate that a person living a standard day-night schedule would be most vulnerable to radiation in the daytime, less so at night. Once this mechanism is thoroughly understood, it may be possible to schedule radiation therapy at a time when cancer cells are at low ebb and most vulnerable, but healthy tissue is in its most resistant phase.[5]

One broad spectrum of human ailments which fluctuate according to circadian rhythms, and similarly respond to drugs and other therapy, is the field of allergies, which cause discomfort ranging from hay fever and hives to bronchial asthma. The latter afflicts some 3 million children, and 9000 people die from it each year in the United States. Allergic reactions may be triggered by plant or animal proteins, drugs, pollutants, house dust, molds, weed and tree pollen, or insecticides—in fact, almost any substance in the environment which may be ingested into the body. Reaction may result from a combination of substances and be intensified by emotional tension. The suffocating symptoms of asthma often strike at night. Recurrence of this disease, or diseases, is related to histamine, a substance which causes a response to insect bites and contributes to the flush in the skin during an allergy attack. An individual's sensitivities often are identified in an elimination process by injecting extracts of various substances to see which will cause a reaction.

In Paris, Dr. Alain Reinberg tested healthy adults by injecting allergenic material at fixed hours day and night. The strongest reaction occurred from doses given at 11 P.M., a time of evening when the corticosteroid adrenal hormones are dropping to their lowest level. Antihistamine drugs were tested in the same way with some persons treated at 7 A.M., others at 7 P.M. The morning dose had a much stronger impact than the evening medication. Also effects of the morning dose lasted seventeen hours, while the evening treatment was effective only a few hours.

Dr. Reinberg also studied patients allergic to penicillin and discovered their greatest irritation occurred at 11 P.M., again coinciding with the low point of adrenal hormones. This repeated evidence of relationship between allergic reaction and the time cycle of the adrenal glands offers a reference point for timing of medication.

Cortisol or cortisone, the adrenal hormones, are given to children to relieve the gasping fits of severe asthma, but these substances have the unfortunate side effect of stunting growth and delaying maturity. At the Children's Asthma Research Institute and Hospital in Denver, physicians tested youngsters with prednisone, a synthetic derivative of cortisone, administered every two hours. Children who received their medication at 1 A.M. or 7 A.M. benefited more than others. Further research may lead to precise ways of choosing the best time of day to administer hormones so they can be given in smaller doses and thus reduce the harmful side effects.

Surgeons have long been puzzled by the fact that some patients recover more quickly than others from identical operations. Though many factors of health and disease enter into this, there is need to time surgical operations during an individual's cyclic peak of vitality. The time of administering anesthetic also may be critical.

Dr. Halberg at Minnesota conducted time series studies with an anesthetic, halothane. He found that ten-minute exposure given at one time of day caused 6 percent mortality in laboratory animals. The same exposure at another time killed 76 percent. Maximum sensitivity occurred in the middle of active time (daytime in the average man), when the animals were least sensitive to other poisons.

Doctors often prescribe drugs to be taken at an easily remembered time, such as lunch or bedtime, rather than taking biological time into consideration. If a drug is known to be more effective or less toxic at a specific phase of the circadian cycle, and this is true of many, physicians should note this in prescribing the best times of day or night to take the medicine. The reason drugs function with different potency at different times may range from oscillations of chemical balance in the brain to the fact that living tissue, even if removed from the body, still exhibits cyclic behavior.[6]

In our frenetic society, in which millions of individuals are dependent upon alcohol and such drugs as stimulants and barbiturates, it should be recognized that a person's reaction and vulnerability to any or all of these may vary widely from one part of the day to another.

One test revealed that if mice were given a measured amount of alcohol at the beginning of the rest period, 12 percent of them died. The same dose given at the time of awakening killed 60 percent of the test animals. True alcoholics may have lost the will to care about such effects, but a person accustomed to social drinking at a certain time, such as cocktails before dinner, should be careful practicing this habit while traveling. He might find himself drinking in a strange land at a moment of vulnerability when the alcohol might trigger illness or death.

Oscillations of susceptibility also should be considered in using sedatives and stimulants. Drs. Lawrence E. Scheving and John Pauly at Louisiana State University Medical Center in New Orleans found that pentobarbital (Nembutal) caused rats to sleep fifty minutes when given at one time of day, but ninety minutes when given at another. Peak response corresponded to the early part of the rodents' activity period.

Similar cyclic differences have been found with stimulant

drugs such as amphetamine. People take stimulants to enhance alertness and energy or to reduce appetite, but it is possible that a dose taken at the wrong time can do serious harm. Since amphetamines are addictive drugs, and popular with youthful drug abusers, knowledge of biological time may assist therapists as they help addicts taper off their harmful habits. The knowledge will be useful to doctors prescribing either sleeping pills for insomniacs or stimulants for people of low energy. It also helps explain why a sedative may induce restful sleep for a person at one time but appear to fail at another.

At the Douglas Hospital in Montreal, Dr. Heinz Lehmann noticed that some of his psychiatric patients were not calmed by the hypnotic drugs they took at bedtime. He experimented by parceling out the doses, giving a fraction in midafternoon, a fraction in the evening, and a final fractional dose before the patients retired. Under this drug schedule, the agitated sufferers were eased into sleep with smaller total amounts of drug than they had been taking before in a single dose.

This promises the day when doctors will be able to achieve their therapeutic goals with smaller quantities of medicine given at the time of greatest effectiveness in synchronization with individual internal clocks. It is possible, too, that by studying their own best times for taking prescribed medication, individuals may gain better effect than by taking a drug on a regular schedule according to the clock. However, at the present stage of knowledge, it would not be wise to experiment without a doctor's advice.

In determining the best times for administering pain-killing drugs, the daily rhythm in human tolerance to pain may furnish a valuable clue. A sick child becomes most uncomfortable and restless at night. At the same time hospital patients demand most pain killers to make them comfortable. Neurologist Henry A. Shenkin at the Episcopal Hospital in Philadelphia theorizes that pain may be related to the level of blood hormones. Since such levels follow a circadian cycle, doctors

Medication Effectiveness

A Guide to Making the Best Use of Your Doctor's Prescriptions

Medication—from aspirin to the most recent wonder drug—is most effective depending upon how "receptive" the body is at any particular time, and the time at which the body is most receptive to any particular drug is likely to be roughly the same every day. Although doctors give their patients detailed instructions when they prescribe drugs (such as: "to be taken every four hours," etc.), they usually do not specify the hours at which the medications should be taken. By deliberately varying the hours at which he takes his medicine, a person can determine those times at which his body is more likely to be receptive and time his dosage accordingly.

Record of Timed Medication Results

Time at which medication taken	Time at which effects are felt	Elapsed time	Check here the shortest elapsed times

Hours at which medication has been
found to be most effective _____

Whenever your doctor prescribes a drug without specifying the hours at which it should be taken, make a chart like the one shown above. Filling it in will tell you at what times the medication is apt to be most potent.

may be able to use blood hormone levels as a measure to distinguish between patients whose pain comes from a physical source and those who suffer emotional or psychosomatic pain.

If pain is periodic, it follows that the time when a pain killer, such as aspirin, is taken helps determine how effective it will be. Researchers found that aspirin taken at 7 A.M. seemed to linger in the body and was detected in the urine twenty-two hours later. If taken at 7 P.M., the drug remained a shorter time and could not be detected after seventeen hours.

Physiological response to pain killers and sedative drugs may be dependent upon rhythms in the nervous system. Poisons, on the other hand, seem to vary in potency according to changes in production of enzymes in the liver. This was shown in animal studies with ouabain, a poison which is used in carefully measured doses, like digitalis, as a heart stimulant and diuretic to increase urine discharge. Test subjects were most vulnerable at the end of their day and most resistant to poison at the end of their rest. The rhythm in liver enzymes, responsible for destroying and removing poisonous substances, also has been established. These cycles are clearly important in determining when dangerous drugs may be given and in what quantities.

. . . How Drugs Desynchronize Body Clocks

IF human biological rhythms affect the potency of drugs, it is logical to expect drugs to have an effect upon those biological rhythms. Disease or trauma causes dissociation of biological cycles. Now investigators finds that the body clocks also can be thrown out of phase by certain drugs.

Antibiotics are a case in point. Actinomycin-D can shift the circadian rhythm in the heart of the living cell, in the synthesis of DNA and RNA molecules which carry the genetic code. Such desynchronization would alter the time at which certain cell populations reach their peak of multiplication and possibly throw this out of balance with other cycles.

Barbiturates, or sleeping pills, also may shift circadian rhythms. Drs. Dorothy and Howard Krieger of Mt. Sinai Hospital in New York City discovered that sodium pentobarbital suppressed the rhythm of adrenal hormones. In another study a short-acting barbiturate blocked the morning rise of hormones in subjects which received it the night before. This effect of sleeping pills upon adrenal hormone rhythms may help explain the "hangover" and mental confusion experienced by many people when they arise in the morning.

The impact of drugs upon internal clocks may be felt far longer than the day they were taken. Dr. Richter at Johns Hopkins gave female rats a germ killer, sulfamerazine. The drug changed their estrus (fertility) rhythm from a normal four or five days to a free-running cycle lasting twenty to thirty-five days. Strangely, the long-term abnormal cycles did not appear until after the drug had been removed.

This caused investigators to suspect that many idiosyncrasies in human behavior and feelings may be the result of aftereffects from medication. Dr. Richter has tested numerous drugs, including thiourea, a thyroid-inhibiting substance; aminopyrine, a pain killer; sodium barbital, a sedative; cortisone, an adrenal hormone, and progresterone, a female hormone. In each case he detected no abnormal effects while test subjects were taking the drugs. Only later, when the drug had been discontinued for some time, many of the animals showed abnormal cycles of eating and activity.

The female hormone, estrogen, left an aftermath of altered cycles in 66.6 percent of animals tested. Since this is one of the hormones commonly used in birth control pills, these tests offer the possibility that some women may experience abnormal physiological and mental rhythms long after they have stopped taking the hormones.

"On the basis of these observations," Dr. Richter said, "the possibility must be considered that the cessation of a drug or hormone after prolonged treatment may also produce lasting effects in man; further, that the existence of such changes may not be detected without the aid of special measurements made over long periods of time." [7]

It is only now that such questions are moving out of the laboratory and coming to the attention of general medical practitioners. As we come to identify the many rhythms and many forces which affect them, perhaps we can begin making good use of our internal time.

Because of the cyclic function of internal clocks, an individual's vulnerability to disease, drugs, poisons, and stress varies from one part of the 24-hour day to another. Awareness of these rhythms of sensitivity and resistance will help physicians and their patients understand previously mystifying individual differences in human health, illness, and response to treatment.

Knowledge concerning these rhythms may:

1. Explain why some people suffer more savage attacks of a disease than others.

2. Assist in the control and cure of drug addicts and alcoholics.

3. Help determine the most effective way to give beneficial drugs and pinpoint times of day when drugs or poison may be dangerous or lethal.

4. Help surgeons to schedule operations according to a patient's cyclic period of greatest vitality and thus enhance his chance for recovery.

5. Serve as a guide for scheduling radiation therapy to take advantage of the time when cancer cells are most vulnerable while healthy tissue is most resistant.

6. Provide more effective control of destructive and disease-bearing insects and microbes by taking advantage of their cycles of vulnerability.

CHAPTER 9

*Different ways of
looking at time.
Psychological and metabolic mapping
to anticipate and relate
to body rhythms.*

Living in Tune—
Discovering
the Beats

WHEN YOU SIT WITH a pretty girl for two hours, you think it's only a minute. But when you sit on a hot stove for a minute, you think it's two hours."

This is how Albert Einstein once tried good-humoredly to clarify his theory of relativity, part of which postulates that time passes more slowly for moving objects than for objects standing comparatively still. This was an abstruse concept in 1905 when Einstein published a thirty-page paper titled "On the Electrodynamics of Moving Bodies," which later became the foundation for his general theory of relativity. Out of it we already have reaped the fledgling world of nuclear power, but it was not until a number of years after the original paper that mathematician Hermann Minkowski deciphered its significance and commented:

"From henceforth, space by itself and time by itself have vanished into the merest shadows and only a kind of blend of the two exists in its own right."

Understanding the space-time relationship will be vital to men of the future as we explore the universe. If the theory is universally true, for example, astronauts would age only four and one-half years during a round-trip journey into space which, according to clocks remaining on earth, lasted thirty-two years.[1]

Most of us will not live to enjoy such a form of time travel or pseudo-immortality, but the theory of time dilation demonstrates how little we understand the nature of time and how to use it. The intelligent human is a paradox. We found it necessary to invent mechanical devices and assign the arbitrary names of years, days, hours, minutes, and seconds to measure the passage of time while inside us, above the level of conscious mind, are many clocks which measure time with great accuracy for the process of living, if we but learn to sense the delicate ticking within us.

Slavery to artificial time often reduces our vitality and highest performance. If we lived by natural time, we might be amazed at our own greater humanity. As J. N. Berrill at McGill University stated it: "Time is of the essence and it is as much a quality of life to measure time and keep in temporal harmony with the spin of the earth and the moon around the sun as it is to grow in space, transform energy, or perpetuate our kind."[2]

As time flows through us—or we flow through it—our flesh and mind, our glands, organs, and cells all are tuned to the cyclic passage of moments, days, weeks, seasons, and years. The rhythms of light and darkness, revolution of the moon around the earth and earth around the sun, and perhaps even planetary motion, all have become engraved within us. Periodicity may have been one of the first forces of natural selection, since organisms that timed their activity and life style in harmony with changing light, temperature, humidity, and other factors of environment would have had an edge on survival. Man's ability to survive is convincing evidence in itself that we developed internal clocks which respond to the day and year, temperature, air pressure, gravity and electromagnetic fields.

... Inner Rhythms in Human Fetuses and Infants

SOME of the rhythms exist even before we are born. Many mothers-to-be in late pregnancy report that junior kicks more lustily at some hours of the day than at others, and studies have shown that periodic activity of the unborn child is even more general than that.

At the Sepulveda Veterans Hospital in Los Angeles, Dr. M. B. Sterman studied pregnant women who volunteered to sleep in the hospital laboratory. They wore electrodes on their abdomens above the fetus and others designed to transmit brain waves, muscle tone, and eye movements. Readings were taken for thirty nights. The data showed two distinct prenatal rhythms in the children. One was an activity rhythm which varied over a 30- to 50-minute time span. The other was an 80- to 100-minute rhythm related to the mothers' cycles of dream sleep. Even though the mothers' sleep was often interrupted during the later weeks of pregnancy, the fetal rhythms continued. After birth, the 90-minute cycle disappeared, indicating it had been induced by the mothers' biochemical oscillations.

Immediately after birth, babies show a 40-minute cycle of breathing, brain waves, and body and eye movements. Their dream cycle in sleep follows a 40- to 47-minute rhythm with waves of sucking, kicking, grimacing, and quiet. By the time a child is about eight months old, the cycle lengthens to ninety minutes again, apparently forming its own independent periodicity with nervous system maturity.

The rhythm of sleep develops gradually during early months, but by the time a baby is sixteen to twenty weeks old it usually has a regular circadian rhythm in its sleep-waking cycle. Other internal rhythms appear according to other time schedules, and bear little relation to a parent's efforts to train the child to the artificial world of adults. By the time an infant is fourteen to twenty weeks old, pulse rate, temperature, urine excretion, and electrolytes such as potassium begin a clear 24-hour oscillation. Other evidence of circadian rhythmicity is not apparent until the child is several years old.[3]

Then, without conscious effort, the human being has joined in harmony with the universe. From that time on, however, we also are forced into a mold of artificial time which dictates hours of sleep, waking, and other activity according to society's patterns. Generally we ignore, or only partially respond to, the natural rhythms of our internal clocks. At the same time —back to Einstein's hot stove and pretty girl—we are puzzled why time seems to pass rapidly or slowly according to different circumstances. This demonstrates that our conscious awareness of time is faulty (in the absence of external signals) and that the only accurate estimates of short time intervals depend upon some internal pacemaker—the biological clocks themselves.

. Time as an Invention of the Human Mind

PSYCHOLOGICAL time, which is what we measure consciously, is a product of intellectual and physical experiences and is shaped by the cultural environment. It is based upon the occurrence and speed of passing events. By our conception, time is given a starting point and eventually an end. We often act as if we were running a race against time; we try to gain time, to do things faster. All cultures have implicit philosophies that are expressed in attitudes toward death, in myths, in language, and in daily conduct, but it is difficult for us to understand that other cultures see time differently from us.

Among the Hindus, the Chinese, and American Hopi Indians, for example, time is experienced as a perpetual recycling. The Hopi language does not contain verbs arranged in present, past, and future tenses. These Indians do not divide time into discrete units that can be added one to another. Rather, they regard time as a repetition of the same event. If days are seen as one recurring event, then how you treat today will have effects on tomorrow. Hopi people emphasize preparation for coming events and believe that by maintaining favorable conditions they can insure the favorable development of later events. Much of their religious ritual is focused on these preparations.

Psychologists, including Dr. Jean Piaget of France, have studied the origin of the sense of time in children and conclude that there is no basic intuition of time within us. At the level of consciousness, the sense of time results only from coordination of speed and distance. In one experiment, Piaget shows a child two small cars. One may travel two feet, the other three. They start at the same time, but the second car travels faster and stops before the first. The child under five thinks that the second car traveled a longer time and he will justify his belief by saying it traveled a longer distance or went faster. Only later can a child combine speed and distance into a unitary concept of time.[4]

This illustrates how time conception is strictly a learning process aided by man-made rather than natural devices. An old Indian, for example, might say that an event took place many moons ago. Our conception of time would translate this into a number of months that had passed. His might be simply that the moon had traveled through that many cycles, without the notion that a *quantity* of time had gone by. The passage of time has been so firmly imprinted in our minds, because of lifelong conditioning, that it is impossible to think of ourselves or our physical and mental health without speculating upon the number of years that are left before death. Is it possible that our lives would be more serene and enriched if we could learn that the years are unimportant so long as our internal clocks retain their natural periodic rhythm?

It is the overlay of artificial time upon our elemental time clocks which often seems to mold our lives into an unsatisfactory pattern. Such patterns would seem more harmonious if we understood our internal cycles. Many things can affect their speed and synchronization. Research at Clark University in Worcester, Massachusetts has shown that time passes more slowly when you're overheated, and more swiftly when you're cold. To a person with a high fever, a few minutes may seem like hours.

Depression or anxiety about the future makes time appear to pass more slowly than it actually does. This is borne out by studies at a Veterans Administration hospital. One depressed patient commented that "getting up this morning seems so long ago that I can barely remember it." Another said: "Yesterday is as remote and far away as events that happened years ago." A third patient, frustrated by the maddening slowness with which time passed, found a revolver and shot up the clocks in his ward.

Elderly people who report that years seem to pass with the rapidity of days or weeks are supported by other studies in VA Mental Hygiene clinics. Age has a definite bearing upon how our internal clocks mesh, or fail to mesh, with artificially measured time. Research data indicates time may seem to pass five times as fast for a man in his sixties as it does for a youth in his early teens.[5]

.. How Artificial Work Schedules Impair Health

OVER a long period of time if a person's natural rhythms persistently run faster or slower than the tempo of his environment, it can lead to mental distress and psychosomatic illness. "Some researchers have observed that the regularity and persistence of stress-producing stimulation seems to be responsible for creating neuroses in animals under experimental conditions," Marc Richelle wrote in *Psychology Today*. "Thus, the arbitrary work schedules that are imposed upon modern man may lead to physical and psychological disorders—as may the deadly regularity of assembly-line jobs."[6]

It may be the same deadly regularity of fixing meals, doing the laundry, and housecleaning which often leads to lethargy and neurosis among housewives whose internal rhythms are plaintively requesting other schedules and other pursuits. At the other extreme are those people who are caught up in the modern world of corporate one-upmanship. There is an almost frantic and perpetual propulsion to higher status, higher sales, higher profits, higher salary, and higher Gross National Product which seems to be the pervading religion in consumer-conscious nations. Many people thrive on this hyperkinetic spiral of super competition which continually compresses more work and creative endeavor into less time. Others do not.

Men and women in this latter group—which may be the majority—are well advised to make periodic, honest surveys of their goals and motivations. Almost everyone knows a friend or acquaintance who is "over his head at the office." That phrase covers a broad spectrum of mental and physical anguish, but it describes the man who is attempting to cope with more work, more pressure, and more frustration than his natural rhythms are able to absorb. A certain amount of work, pressure, and discipline are necessary for human achievement, but the man who forces himself continually to perform above his optimum level of ability and creativeness, perhaps in the pursuit of false goals or out of false pride, is undoubtedly shortening his life. Persistently working out of phase or beyond the tempo of individual biological clocks may contribute more than we know to the rising incidence of stomach ulcers, mental breakdown, and fatal heart attacks.

. . . How to
Tune in Your Rhythms

ON the positive side, there are a number of things you can do—based upon current scientific knowledge—to take advantage of your natural rhythms and make better use of time:

1. Become conscious of individual cycles in mood and physical strength, and take these into consideration in daily living.
2. Help the body to adjust if a major phase shift or stressful disruption of normal rhythms should occur.
3. Determine, at the medical level, the most opportune time—as shown by individual measurements—for taking drugs and undergoing therapy such as X-ray treatment or surgery.

Studying your own individual rhythms, charting recurring

high and low periods and changing habits accordingly, can be richly rewarding. It is not yet possible for the average person to have a computer map drawn of his myriad cycles of temperature, hormones, urinary excretions, and blood sugar level, but you can learn much by keeping simple records over a period of time of how you feel during certain hours of the day, days of the week, months, and seasons.

Possibly the most important cycle is sleep, which also is easiest to measure in terms of your own needs and best rhythm. Many people function well with only four or five hours per night, while others need eight or more. Some people do their best work at night, while others are most alert and productive in the early morning. These preferences often are subordinated to the need not to disturb others by keeping "odd" hours or the need to conform to society's nine-to-five work day. Therefore members of an average family habitually go to bed and get up in the morning at approximately the same hours.

The body's own rhythms of strength and weakness are trying to signal the proper time for sleep and wakefulness. A person who is internally synchronized to a 23-hour or 25-hour day, rather than an exact 24-hour circadian rhythm, if allowed to run free might obtain his best performance and well-being by going to bed an hour earlier or later each night. Inevitably, of course, this would eventuate in periods during which hours of work and sleep would overlap day and night. This would be intolerable for a person locked into a nine-to-five job. Such a man's efficiency might improve greatly if it were possible for him to work according to his natural cycles. Another example is a child who is forced to go to bed every night at eight o'clock. He might function better and improve in mood and behavior if allowed to sleep when he felt tired. Unfortunately, as any mother knows, this is not possible when the child must meet the 8 A.M. bus for school next morning.

Many people who suffer from insomnia are victims of the same artificial schedule keeping. A woman who knows she needs eight hours of sleep in order to cope with tomorrow's problems may retire at 10 P.M., then toss and turn until midnight, when her internal clock decrees it is time to sleep. Over the years such a forced pattern becomes a habit of insomnia. Most people in modern society control it with sleeping pills. The drugs in turn may throw the internal rhythms out of phase and lead to the commonly experienced sleeping-pill hangover. There are various reasons why a person may be unable to sleep, but many insomniacs might cure themselves by studying their own periodicities, then sleeping and waking accordingly. It might throw the household schedule out of kilter, but some insomniac housewives might make life more cheerful for themselves and their families if they followed such advice.

Fortunately, this advice can be followed without resort to complex and expensive medical tests. A person's best time for sleep and activity follows the temperature cycle, which falls one or two degrees in the evening and then rises to normal (approximately 98.6° F.) shortly before waking in the morning. The times at which the drop and rise occur may vary by as much as several hours from one person to another. The man or woman who leaps out of bed full of vim and vigor in the morning usually is one whose body temperature drops steeply at night and rises to normal an hour before waking. Such a person also falls asleep easily. The opposite is the person who falls asleep with difficulty and then wanders around at low ebb for two or three hours after getting up in the morning before reaching normal levels of efficiency. This individual probably will find that his temperature (an indirect measure of metabolism) drops slowly at night and does not rise to normal until long after the alarm clock rouses him from bed. The slow riser might function better if he retired later and rose later.

Your best cycle may be determined by taking your temperature in early morning, at midday, before dinner, and just before sleep. Hourly temperature readings would provide an even more accurate chart. This might also show a small dip in temperature in afternoon or evening, a signal that a nap might be highly beneficial.[7]

The same temperature chart also may help identify your best hours of physical fitness and work efficiency during the day. This not only would enhance physical and mental performance but also help eliminate the concern and anxiety which most people feel when they are at low ebb. If they understood this is a natural part of their normal cycle, they could take it into rational consideration as part of work planning. Such knowledge, and consideration, is especially important for those in demanding jobs which require precise performance and high levels of alertness, such as airline pilots or air traffic controllers. Metabolic maps of internal rhythms soon may be recognized as essential measurements in the examination of people who are to be placed in such jobs.[8]

Ideally, individual time maps then would be followed in scheduling work shifts. It is possible someday that the schedulers in American commerce and industry may recognize the importance of optimum sleep-activity cycles in individuals and assign them to work accordingly. This would create a revolution in the frozen seven- or eight-hour shift schedules of industry, but it could pay dividends in greater human productivity. If everyone could work on individually staggered schedules, it also might help to alleviate the massive traffic jams which clog metropolitan centers every morning and evening.

While waiting for that day of enlightenment, you can chart some of your internal cycles and live by them more completely, even within the confines of social pressures. Self-understanding begins with careful notes of standard activities as they fluctuate during the day, and eventually over longer cycles of weeks and months. Records should range from sleep habits to eating and defecation, sensory alertness, and mood shifts. This way you can build your own metabolic time map, crude though it may be, without laboratory measurements and computer analysis.

One way to start is to rule a pad of paper vertically into twenty-four hourly segments and check off the hours when life events usually happen, or when you feel they should happen. The previously mentioned temperature cycle may be the key to others. Another is the hour when you normally feel sleepy, as well as the hour you would get up in the morning if allowed to sleep through the alarm clock. (A series of weekends should provide this record even if job or school dictates early rising on weekdays.)

When charting your sleep patterns, remember everyone usually sleeps in repeating cycles of 90 to 120 minutes each, with dreams near the end of each period. If you frequently remember that you dreamed, even if the content of the dream is not recalled, it is an indication that your sleep cycles are normal. You may find that you smoke heavily, eat snacks, or daydream at certain regular times each day. Awareness of such cycles may help you enjoy each day more and also begin training yourself to better control, synchronizing the conscious sense of time with the internal clocks. Such control is what some people exhibit instinctively when they keep precise appointments or waken in the morning at the minute they choose.

Another line on the chart may measure the sharpness of your senses. Because of the rhythm of the carbohydrate-active hormones from the adrenal glands, your senses are sharper at some times than others. In most people, this occurs in the evening when the adrenal hormones are at lowest ebb, but here again the cycle may vary considerably from one person to another. Once the sensory rhythm is established, you may be able to enjoy favorite pleasures more keenly.

This is one reason why a heavy evening dinner is most enjoyable, but the pattern of eating should be established against other indications that proteins are utilized by the body most efficiently in the morning. This could be a guide, especially for weight watchers, that the heaviest meal should be in the morning, with a smaller second meal and a light supper.

Your chart also should include, as objectively as possible, hours of daily mood change. Though love-making should be a romantic business not calculated against the clock, awareness of predictable moods and sensory keenness may help a woman guide her husband to the most propitious moment—or vice versa. At the same time, the male glands have something to say in the matter. Usually the male hormone testosterone reaches its cyclic peak shortly after the hour of awakening in the morning. Also, most men awaken with an erection which is related to their last dream period.[9] Charting such rhythms, and remaining aware of their physiological foundation and likely recurrence, could go far toward eliminating some of the boredom and friction in many marriages.

In basic health habits, charting your normal hour of defecation also is important. Irregularities in this function often are an accurate indicator of approaching illness.

When daily ups, downs, and idiosyncrasies have been mapped, your chart probably also will indicate longer trends of mood and physical fitness. A person may feel great one day but lousy the next, or he may find the bad days grouped in a certain part of a several-week cycle. Such rhythms are more difficult to delineate because feelings of happiness or depression seem most often to be governed by external events such as financial success, failure to reach goals, or a fight with one's wife. However, the longer cycles do exist, especially in the female menstrual cycle. Once a woman is aware of the fact that she may suffer several days of premenstrual tension, conscious effort can help avoid family friction and disagree-

Mood Chart
A Guide to Activity Planning

Here's how to discover your characteristic mood cycle, i.e., the number of days on which you feel good, followed by the number of days on which you feel bad, which recur in a comparatively constant and fairly predictable pattern. First, draw a calendar chart like the one on the opposite page. Then, over a period of as many months as it may take to determine a comparatively constant cycle, fill in the boxes representing the days during which you are elated, happy, calm, etc. with a G (for "good"); use a B (for "bad") to indicate the days on which you feel generally depressed, unhappy, irritable, etc. Many people find a variation of up to two days in the duration of the periods during which they feel bad or good. Within a few months, a fairly regular pattern of Gs and Bs will emerge. There may be only two or three Gs or Bs in a row; on the other hand, there could be five or more boxes of one letter in sequence. Once you have determined this recurring pattern, mark your own daily pocket or desk calendar in any of a variety of ways to flag the probable onset of "good" and "bad" periods. Such a guide will help you to undertake difficult tasks at times which minimize the possibility of failure, warn you when prudence is advisable in handling romantic interests, and assist you in scheduling the activities you enjoy most on those days when you will be most likely to enjoy them best.

Sample Mood Chart
(PARTIALLY COMPLETED)

JAN 1	2	3	4	5	6	7	8	9	10	11	12	13	14	15	16	17	18
19	20	21	22	23	24	25	26	27	28	29	30	31	FEB 1	2	3	4	5
6	7	8	9	10	11	12	13	14	15	16	17	18	19	20	21	22	23
24	25	26	27	28	MAR 1	2	3	4	5	6	7	8	9	10	11	12	13
14	15	16	17	18	19	20	21	22	23	24	25	26	27	28	29	30	31
APR 1	2	3	4	5	6	G 7	G 8	G 9	G 10	G 11	G 12	G 13	B 14	B 15	B 16	G 17	G 18
G 19	G 20	G 21	G 22	B 23	B 24	B 25	G 26	G 27	G 28	G 29	G 30	G 1	G 2	G 3	B 4	B 5	G 6
G 7	G 8	G 9	G 10	G 11	12	13	14	15	16	17	18	19	20	21	22	23	24
25	26	27	28	29	30	31	JUN 1	2	3	4	5	6	7	8	9	10	11
12	13	14	15	16	17	18	19	20	21	22	23	24	25	26	27	28	29
30	JUL 1	2	3	4	5	6	7	8	9	10	11	12	13	14	15	16	17
18	19	20	21	22	23	24	25	26	27	28	29	30	31	AUG 1	2	3	4
5	6	7	8	9	10	11	12	13	14	15	16	17	18	19	20	21	22
23	24	25	26	27	28	29	30	31	SEP 1	2	3	4	5	6	7	8	9
10	11	12	13	14	15	16	17	18	19	20	21	22	23	24	25	26	27
28	29	30	OCT 1	2	3	4	5	6	7	8	9	10	11	12	13	14	15
16	17	18	19	20	21	22	23	24	25	26	27	28	29	30	31	NOV 1	2
3	4	5	6	7	8	9	10	11	12	13	14	15	16	17	18	19	20
21	22	23	24	25	26	27	28	29	30	DEC 1	2	3	4	5	6	7	8
9	10	11	12	13	14	15	16	17	18	19	20	21	22	23	24	25	26
27	28	29	30	31													

The above chart contains all 365 days of the year in consecutive order. The person filling in the chart has marked those days on which he felt good with a "G" and those days on which he felt bad with a "B." Although the chart has been only partially completed, a pattern is already beginning to emerge. It shows recurring cycles of from 6 to 8 "good" days followed by 2 to 3 "bad" days. This information enables the person completing the chart to schedule his most difficult jobs and most enjoyable activities on those days during which he is most likely to feel good and his easiest, most routine jobs and least enjoyable activities on those days during which he is most likely to feel bad.

ments which are intensified during periods of depression. The husband's rhythms may not be so evident, but once determined he, too, can consider this as much a part of his life as the weekly pay check. By understanding—and respecting—each other's cycles, a husband and wife will find greater harmony in living by making the most of upbeat days and avoiding quarrelsome subjects on the bad.

These basic steps toward living most effectively with your natural rhythms may seem simplistic, but the most difficult task will be that of identifying the subtle oscillations which guide individual lives, either in harmony or conflict with artificial schedules. Knowing yourself more profoundly can lead to a richer, more productive life.

Artists, writers, musicians, salesmen, and others who are not bound by strict work schedules can put this knowledge to work directly. Those people who find themselves persistently out of kilter with their professional lives may find great advantage in reassessing goals and motivations with a view toward changing their profession to something more humanly rewarding even if the economic returns are less. At the very least, knowledge that good and bad periods in the cycle are normal to everyone can help erase the burden of worry which accompanies depression and low periods, and perhaps reduce psychosomatic illnesses which arise from emotional tensions.

. . . How to Assist the Body in Phase Adjustments

ONCE you have established a reliable chart of your biological rhythms, the next step in harmonious living is to help your body and brain recover from the effects of an event which throws cycles out of phase. You may find that a bad day may occur about forty-eight hours after an all-night party, a change in work shift, an east-west flight, emotional blowup, or after taking sleeping pills. Depressant drugs appear to shift the daily adrenal rhythm, which may account for the delayed hangover.

As for sharper phase shifts, previous chapters have listed steps which will alleviate upset and illness following long jet flights across time zones or a change in work schedules. Various rhythms of the body require shorter or longer times to adjust for such shifts.

Because various drugs and other substances have a more powerful impact upon the system at one time of day than another, the traveler who regularly takes prescription drugs should be careful. If his stay in a strange land is to be short, he should continue taking the prescription at the same hours he did at home. If the trip is a vacation more than a week long, he should shift the hours of drug-taking gradually. The same advice holds true with heavy eating and drinking at odd hours. Such a careless regime could trigger a heart attack or other severe illness.

At the medical level, general practitioners are just starting to take biological time into consideration when prescribing therapy, drugs, or surgery. This is because other symptoms and factors are more obvious and require immediate attention without consideration of a patient's internal time. Also, most doctors do not have the time or equipment to build individual metabolic maps which may become a standard part of medical practice in the future. In the meantime, if a person becomes reliably aware of his own internal periodicities, he can call this to his doctor's attention and perhaps space the taking of drugs at the time when they will do the most good.

We are only at the beginning of a time when the biological clocks will give us reliable clues of approaching illness, but as stated by Dr. Bertram Brown, director of the National Institute of Mental Health:

"Not knowing that one has a time structure is like not knowing one has a heart or lungs. In every aspect of our physiology and lives, it becomes clear that we are made of the order we call time. As we look deeper into the dimensions of our being, we may find that we, too, are like the plant that flowers if given a little light at the right time every seventy-two hours.

"There may be in man a combination lock to his activity and rest, his moods, illnesses and productiveness. Moreover, by cultivation, he may learn to utilize his subjective sense of time." [10]

Thus, by drugs and other means, there now is a hint that knowledge of internal rhythms may bring the ability to manipulate time and thus add a new dimension to our future.

Time is relative, not absolute. Time seems to move fast or slowly according to whether a person is hot or cold, happy or depressed, young or old. When the tempo of the environment is faster or slower than one's internal clocks, mental distress and psychosomatic illness may result. That is why arbitrary work schedules can wreak havoc with a person's health and why many people who find their schedules too stressful should seriously consider changing their line of work.

By keeping track of his daily changes in body temperature, hours of eating and defecation, mood shifts, and sensory alertness at different times of the day, an individual can make his own metabolic time map which will enable him to live by the schedule which suits him best.

CHAPTER 10

New vistas in human self-control.
Meditative ways
to alter rhythms of the body
and treat illness.

Can Man Affect His Inner Clocks?

LOOKING INTO THE FUTURE of biochronological research is a tenuous undertaking because the past is so new.

Information gleaned in the past ten years, from study and measurements not possible twenty years ago, still has not coalesced firmly within the pristine framework of objective science. Almost everywhere in nature we find evidence of internal clocks at work, keying bugs, plants, animals, and people to the cyclic rhythms of the universe. Yet Western science, with its obsession for quantitative measurement, is not satisfied until the locus—the finite center—of all this marvelous automatic coordination can be identified and established once and for all in a specific place. Such a specific place or object may not exist.

It is both frustrating and exciting that biological clocks can be found everywhere from glands to individual cells, while the windup key for their precise but changeable operation is as elusive as angels dancing on a dust mote. The key or keys to the clocks may range from the shape of certain hereditary cells to a spectrum of unmeasured forces flowing from far-away planets.

But in our age, which demands precise dimensions and physical properties for every phenomenon in nature, it also is the nature of man to manipulate whatever he finds. The very diffusion of the biological mechanisms may force investigation farther afield and eventually link Western science to transcendental forces and techniques which hitherto have been rejected as myth or falsehood.

The growing belief that the clocks are endogenous—born within us—gained one more piece of supportive evidence late in 1971. Theorizing that mutation of certain genes might lead to abnormal rhythms, biologists Ronald Konopka and Seymour Benzer at the California Institute of Technology in Pasadena exposed fruit flies to a drug known to cause mutations. The two found that altering a single gene changed the periods of insect activity as well as the 24-hour cycle in which adult flies emerge from the pupae. Though not final proof that the key to the clocks is internal, these experiments did show that the genes, which carry all of our hereditary information, play a specific role in determining or specifying the rhythms of life.[1]

"It seems that the clock must reside in some region of the brain functionally distinct from that responsible for awareness of time," stated Dr. Mills of the University of Manchester. "Many functions show circadian rhythmicity but this is often merely impressed by external rhythm or habit of environment.

"There is, however, a circadian clock which may be placed tentatively in the region of the hypothalamus, influencing a variety of functions through many channels, known and unknown, and it may itself be influenced by various environmental stimuli.

"It is perfectly conceivable," Dr. Mills added, "that endogenous rhythmicity is present at many levels of organization. Each would normally be entrained by another rhythm, external or internal, thus securing the customary integration and synchronization of different functions, but leaving varied possibilities for disturbance that have hardly yet been explored in man." [2]

... Changing Biological Rhythms with Drugs

EXPLORING the disturbance of internal rhythms is the primary task ahead. We have seen the evidence of harm which can come if our biological clocks are inadvertently disturbed. And we have seen some ways in which we can tune in to our own rhythms and use them for a more serene and productive life. The next step, now unfolding, is to learn how to shift the rhythms deliberately when it is desirable to do so. One possibility is the use of drugs or timed administration of the body's own extracts.

In Australia a drug has been developed which controls the breeding cycle in sheep. Extension of this line of research to other farm animals would create great economic benefit by allowing farmers to breed cattle and market their livestock at the most profitable times.[3] At the human level, such technology could extend the range of drugs and devices available for selective population control.

Since many drugs are known to cause phase shifts in certain body rhythms, we may find a drug which will help us adjust more rapidly after a shift in work schedule or jet flight. The Syntex Corporation, in its investigations with Trans World Airlines, is attempting to determine if certain adrenal hormones, given to travelers, might speed up their phase shifts and make adjustment easier to new time zones.

Another substance is being investigated by Drs. Michel Jouvet and Jacques Mouret of the School of Medicine in Lyon, France. This is a class of drugs known as monoamine oxidase inhibitors. Monoamine oxidase is an enzyme which breaks down and prevents accumulation of certain brain substances known as monoamines. Drugs which inhibit action of this enzyme have been used to ease depression in human patients. The French investigators have found also that monoamine oxidase inhibitors can change the activity-sleep cycle in rats. They feel that in humans it could be used to shift the 24-hour cycle of sleep and waking to as long as forty-eight to seventy-two hours.[4]

. . . Changing Biological Rhythms with Meditation and Biofeedback

As shown by Dr. Janet Harker's experiments with cockroaches, if the seat of the biological clocks is ever discovered in man, it might then also be possible to shift our rhythms selectively by surgery. However, both drugs and surgery essentially represent the brute force phase of learning to control human physiology and behavior. Now some psychologists, including Dr. Neal E. Miller of Rockefeller University, believe it may become possible for people to train themselves to control their own inner functions. These may include the biological clocks and other internal rhythms such as hearbeat and breathing, which are controlled by the autonomic nervous system and have long been considered to be beyond reach of conscious control.

It is here that Western science begins to cross paths with

esoteric disciplines long practiced by religious mystics of the East. It begins, perhaps, with *yoga*. This is a Sanskrit word which names the school of philosophical Hinduism in which it is believed that union with Brahman, the Absolute Being or World Soul, can be achieved through mental and physical discipline. Yoga offers detailed directions for suppressing internal body activity, including breathing. Mental activity also is suppressed until the individual comes into a state of blissful, serene contemplation of Brahman.[5]

Yoga, as taught by Maharishi Mahesh Yogi, who influenced the Beatles for a time, has become one of the many fads indulged by Western youth in their scattered efforts to achieve higher states of physical and mental awareness. It coincides with another fad, or budding science, whichever the case may prove to be, of modifying and controlling cyclic brain waves with the same goal of reaching a state of calmness and serenity.

Beginning as early as 1964, psychological experimenters found, through electronic brain wave measurement, that people can be trained to produce alpha brain rhythms, which oscillate at about ten cycles per second. Such rhythm normally occurs when the eyes are closed and accompanies feelings of relaxation and serenity. This state also renders a person more receptive to deep meditation.

By amplifying tiny electric impulses to show a person what is happening in his autonomic nervous system—a process known as biofeedback—an individual may be trained to control automatic functions including heart rate, blood pressure, or even glandular activity.

Dr. Miller, one of the most successful experimenters in this esoteric field, has obtained almost phenomenal results with trained laboratory animals. Rats have been taught to change the frequency of intestinal contractions, the filtration rate of their kidneys, and the amount of blood flowing through stomach walls, tail, and even their ears, and to alter their blood pressure. In one case the training was so precise that a rat could be induced to blush with one ear but not with the other.

A thirty-three-year-old woman who suffered hypertension (high blood pressure without an obvious cause) was trained to raise or lower her blood pressure at will. The training consisted of monitoring her blood pressure continually and rewarding her immediately—with a sound or light signal—whenever her blood pressure fluctuated in the desired direction.[6]

The limited experience so far obtained in biofeedback training indicates that not everyone is susceptible of achieving this relaxed, highly aware state denoted by alpha brain waves. However, a number of important laboratory clinics now are teaching many patients to control heart rate, lower blood pressure, change skin temperature, relax certain muscles, reduce a compulsion to eat, and enhance memory and creativity. This enables them to control—without drugs—such ailments as cardiac arrhythmias, insomnia, migraine and tension headaches. At the Menninger Foundation in Topeka, Kansas, people have learned to prevent headaches by mentally altering the temperature and blood flow in their foreheads and hands.[7]

To the exploring young, biofeedback training has been dubbed an "electronic guru." One experimenter compared it with flavor enrichers in food. "Biofeedback training is to the mind what monosodium glutamate is to food," he commented.

At California State College in Fullerton some seventy students now are working regularly with brain wave feedback devices.

"I used to get very nervous just before exams," related Steve Nowacki, one of the students. "Now I simply go into alpha about half an hour before taking tests and it gives me a certain alert relaxedness.

"As a meditator, I've been sort of ho-hum, so-so. Meditation puts you into the guru game. You start following somebody and absorbing philosophies and theologies. Something always turned me off with every guru I came in contact with. It's different with an electronic guru."

Others describe the effects of brain wave manipulation as similar to the experience of mood-altering drugs—without the hangover and danger of brain damage. Roy Tuckman, whose chief interest is music, reported satisfying results with biofeedback training.

"The music I improvise on my banjo in that state of consciousness is so new, sometimes pretty wild and pretty nice," he said.[8]

The striking similarity between these mechanical psychological techniques and the transcendental meditation of yoga motivated Harvard University researchers to measure the physical and mental changes which occur during such intense meditation. Dr. Herbert Benson and R. Keith Wallace reported that subjects in such a withdrawn "trance" showed a slower heartbeat. Electrical resistance of the skin increased, indicating that the person was relaxed. The subject's body produced smaller amounts of carbon dioxide and brain alpha waves increased in intensity, another sign of relaxation. Less lactic acid was produced in the blood, a possible indication of reduced anxiety.[9]

Such studies indicate a striking agreement between the newly discovered techniques of biofeedback control and the ancient art of transcendental meditation in offering a greater ability to control our own internal rhythms and functions—a true measure of the possibilities in the mind's control over the body. Other benefits of meditative techniques were discovered by Dr. Benson's study of 1862 youthful drug abusers who also had tried transcendental meditation for at least three months. In almost all cases—nineteen out of twenty—the young people reported they had given up drugs because their subjective experience in meditation was superior to the effects produced by drugs. In this way, both biofeedback and meditation may prove useful tools in curing addicts, alcoholics, psychosomatic illness, and many neurotic compulsions.

The newfound ability to self-train the body's unconscious functions, known to religious mystics for centuries, also may help to explain some of the deepest personal religious experiences of the West, particularly the stigmata.

Probably the most famous example of this was Saint Francis of Assisi who, while in meditation and prayer in the year 1224, experienced the wounds of Christ. His hands and feet were perforated and a lance wound, fresh and bleeding, was seen on his side. Blood from this wound, then and later, actually stained his garments.

Of more recent notoriety was the Austrian nun Theresa Neumann, who was born on Good Friday, April 8, 1898 and first received the stigmata on Good Friday in 1926. Through the years, thousands of people witnessed recurring nail wounds in her hands and, as she meditated on Christ's suffering, her eyes also bled profusely.[10]

Is it possible, through intense meditation upon the manner of Christ's crucifixion, that these people unknowingly were able to induce their internal organs and processes—their very flesh—to reproduce the wounds of Christ?

It may be just as well to leave such matters in the realm of the miraculous for the time being, but the ability by meditation or training to gain control of internal processes and rhythms offers great hope for alleviating many human ailments and distress. It also may help to lengthen that most important cycle of all, the years of our lives.

. . . How Control of Body Temperature May Lengthen Life

ALTHOUGH thousands of factors from birth to death contribute to how fast we age and die, one is temperature, as shown by ingenious experiments conducted by Dr. Charles Barrows, Jr. He is a member of the staff of the Gerontology Research Center directed by Dr. Nathan W. Shock in Baltimore. Barrows's work adds one more clue to link biological rhythms, biofeedback, and transcendental meditation with the problem of aging.

In preceding chapters it has been shown that the temperature cycle is one of the most stable biological clocks in the human body and an indicator of rhythm in other metabolic processes. In biofeedback research, patients have been trained to control their temperatures selectively.

In Baltimore, Dr. Barrows worked with a tiny aquatic animal known as the rotifer. Only half a millimeter in length, the rotifer is parthenogenetic (meaning self-fertilizing; thus many identical specimens may be obtained from the same egg) and its body temperature is always the same as its surroundings.

Dr. Barrows found that rotifers grown in water at 35° C. lived only eighteen days. When the temperature was reduced to 25° C., they lived thirty-four days—almost double their normal life-span. Testing rotifers with combinations of diet and temperature also indicated that most of the life gain occurred in their younger reproductive "years."

What does this mean for man?

Dr. Bernard L. Strehler, director of the biological laboratory at the University of Southern California's Rossmoor-Cortese Institute for the Study of Retirement and Aging, sees the possibility that cooling human temperature by one or two degrees could add up to twenty years to normal life expectancy. Drugs to accomplish such a temperature change are under experimentation, but if we can learn to control our own temperature rhythms, the knowledge of our biological clocks may help us to live longer as well as richer lives.

"The rotifer experiments suggest that if there are two discrete parts of the life program," Dr. Barrows said, "procedures might be developed to interrupt or slow down the aging process, either in youth or after one has reached adulthood. What those procedures might turn out to be is, at this point, anybody's guess." [11]

One procedure could be another rhythm-related mechanism which someday may permit human beings to hibernate as do many lower animals. James M. Lyons and John K. Raison at the University of California at Riverside found that the mix of saturated and unsaturated fats in the membranes of individual cells determines the temperature below which life processes come to a stop. The mix of these fats, according to the investigators, is determined in some way by the periodic cycles of working parts within the cell.[12] There are few conditions under which the average person might wish to retire into hibernation for long periods of time, but such slowing of the life cycle also could extend the total years of life. It also could be important when man in his space quest ventures beyond the solar system in interstellar trips of many years' duration.

So the microscopes and mathematics of science, despite the mundane quarrels and unrest of society, keep pointing our eyes to a future in which we can become better synchronized with ourselves and the universe.

Now that we know we are children of time, it is only a matter of time until we learn to control it—and ourselves.

> Now that man has found ways to live a healthier, happier life by tuning in to his biological rhythms, he may discover how to change them deliberately through the use of drugs, surgery, the meditative techniques of Yoga, and biofeedback. Already, people have been taught in the laboratory to produce alpha brain waves, raise or lower their blood pressure, and control their blood flow. Someday people may be able to lengthen their lives by controlling their body temperature.

Ten Ways
To Use Your
Biological Clocks

Creativity. Because of shifting internal rhythms, some people are
more creative by day, others by night. Chart your own most
brilliant hours of the day, or days in the month, and use the
chart as a guide for the future.

Drugs. Many drugs are more potent at one time of day than another.
By studying your rhythm, you can determine when to take
aspirin or sleeping pills, for example, for maximum effect with
smallest dose. Always follow your doctor's advice where any
drugs are concerned.

Eating Habits. The average person's sense of taste and smell is
keenest in the evening, making dinner most enjoyable. However,

you may make better use of protein eaten in the morning. A dieter may thus obtain best results with a heavy breakfast, moderate lunch, and light supper.

Insomnia. Best times for sleeping are indicated by body temperature, which usually falls in the evening and rises in the morning. Some insomniacs go to bed too early. If you sleep poorly, without obvious cause, chart temperatures and go to bed when your internal rhythms say it's time. A new, more satisfying sleep habit may result.

Making Love. All senses are keenest for most people in the evening. However, male sex hormones reach their peak in the morning. Combine these facts with a chart of your (and your wife or husband's) moods to choose times for making love and enhance pleasure.

Mood. Almost everyone's mood fluctuates during the day as well as during the week or month. Chart yours to avoid arguments on bad days and make the best of good days. In most women, days of premenstrual tension and depression should be clearly known and activities planned accordingly.

Sleep. Some people function best by going to bed early and rising early, others by retiring late and rising late. Some need more sleep than others. Again, the temperature cycle can help you find out if you're an early bird or an owl. Recalling periods of dreaming is an indication that sleep cycles are normal.

Travel. Rapid travel across time zones can upset internal rhythms. If you plan a distant trip by jet, be rested before you start and begin advance shifting of sleep and eating habits toward the time zone of your destination. Once there, do not overindulge in food and drink until your system has had time to readjust. Be gradual in shifting the times for taking prescription drugs. Allow time for rest before demanding activities.

Work Planning. Physical fitness, efficiency, and mood all vary cyclically during the day and fluctuate over longer periods of time, up to six weeks in the average person. Chart your moods and feelings and use this information to plan most demanding work on good days, routine on the bad.

Work Shift. Changing a work shift from day to night or vice versa shifts the phase of internal rhythms in the same way as jet flight. Taking similar measures to pamper your cycles during the period of readjustment can reduce discomfort and enhance efficiency on the job.

Glossary

acidosis—decreased alkalinity of the blood and tissues, usually the result of excessive acid production.

adrenal glands—two endocrine glands near the kidney, producing adrenaline and several hormones which control salt and water balance, sodium and potassium metabolism, and use of glucose and certain steroids related to sex hormones.

biochronology—the study of time relationships with biological processes.

carbohydrate—a class of food including sugars, starches, dextrans, glycogens, and cellulose.

chronopathology—the biological time structure of illness.

circadian—from the Latin *circa diem*, meaning around a day.

corticosterone—one of several steroid hormones extracted from the adrenal cortex.

DNA—deoxyribonucleic acid, the carrier of genetic information in the nucleus of cells.

diuretic—a substance which increases the flow of urine.

diurnal—used here to describe creatures active by day, including man.

edema—abnormal accumulation of fluid in tissue, causing puffy swelling.

electroencephalograph—an apparatus for detecting and recording brain waves. Electroencephalogram is the tracing.

electrolyte—a substance, such as acid, base, or salt, which, when dissolved or fused, becomes an ionic conductor.

endocrine system—internal ductless glands which secrete hormones.

endogenous—originating within the body.

enzyme—one of many proteinaceous substances, produced in living cells, that act as catalysts in promoting biochemical reactions at higher speed than normally would be achieved at body temperatures.

epinephrine—the principal blood-pressure raising hormone from the medulla of the adrenal glands; used synthetically as a heart stimulant.

estrus—regular recurring state of sexual excitability in the female of most animals. Time of fertility.

exogenous—originating from outside the body.

gamma globulin—serum containing antibodies against disease.

geriatric—related to aging or the process of aging.

gerontology—scientific study of aging and the aged.

glucagon—crystalline protein obtained from the pancreas. Assists the breakdown of glycogen in the liver.

glucose—a sugar normally found in the blood; chief source of protoplasmic energy, the form in which carbohydrate is assimilated into the animal body.

glycogen—a sugar, the principal form in which carbohydrate is stored in animal tissue, especially in liver and muscle.

homeostasis—used here as a tendency toward stable environment in bodies of higher animals through interacting physiological processes.

hormone—an organic product of living cells which affects activity of cells remote from its point of origin, such as secretions of the endocrine glands.

hyperventilation—excessive respiration, leading to abnormal loss of carbon dioxide from the blood.

hypochondria—depression centered upon imaginary physical ailments.

hypothalamus—a part of the brain important in regulating autonomic (involuntary) body functions.

infradian—in this context, denotes a cycle or rhythm longer than one day in period.

insulin—a pancreatic hormone necessary for the metabolism of carbohydrates.

LSD—lysergic acid diethylamide, a hallucinatory drug.

mescaline—a hallucinatory drug obtained from mescal cactus.

microfilariae—microscopic larvae of certain parasites, such as those causing filariasis (elephantiasis) in humans.

mitosis—division of the living cell.

neurosis—a disorder of the central nervous system usually manifested by anxiety, phobia, obsession, or compulsion.

nocturnal—used here to describe creatures active by night.

norepinephrine—a hormone occurring with epinephrine; has strong action in constricting blood vessels and affects transmission of nerve impulses.

nymphomania—excessive desire by the female for sexual activity; usually rooted in feelings of inadequacy.

oncology—the study of tumors.

ouabain—a toxic crystalline steroid obtained from seeds of an African shrub.

pancreas—a gland lying behind the stomach; secretes digestive substances and insulin.

parasite—an organism living in or upon another organism.

parthenogenesis—reproduction without need of fertilization.

phase shift—used here as the condition when a body cycle or rhythm is forced into, or adjusts to, a new time frame.

pineal gland—an endocrine organ in the brain, sensitive to light and sometimes considered a vestigial third eye. The pineal is important in the regulation of some body rhythms.

pituitary gland—an endocrine organ attached to the brain; helps to regulate other glands along with other body functions.

plasma—the fluid part of blood or lymph.

protein—a combination of amino acids; essential constituent of all living cells.

psychosis—profound disorganization of mind, personality, or behavior, usually stemming from inability to cope with the demands of the social environment.

psychosomatic—physical symptoms resulting from emotional conflict; often erroneously termed "imaginary illness."

pulmonary—relating to, or associated with, the lungs.

REM—Rapid Eye Movement in sleep, usually associated with dreaming.

RNA—ribonucleic acid, a hereditary message bearer in the living cell.

steroid—a class of compounds with polycyclic structure; includes vitamin D, bile acids, and various hormones.

substrate—the base upon which an organism lives.

triglyceride—a tri-ester of glycerol with one, two, or three acids. Occurs in natural fats.

ultradian—biological cycles occurring in intervals shorter than one day.

vector—an agent, such as the housefly, which carries a disease from one creature to another.

zeitgeber—a German word denoting the force which entrains, or triggers, a biological rhythm into action.

Notes

Notes to Introduction

1. *Biological Rhythms in Psychiatry and Medicine* (Washington: National Institute of Mental Health, 1970), p. iv.

Notes to Chapter 1

1. Michel Gauquelin, *The Cosmic Clocks* (1967; reprint ed., New York: Avon Books, 1969), p. 47.
2. Ibid., p. 177.
3. Ritchie R. Ward, *The Living Clocks* (New York: Alfred A. Knopf, 1971), p. 10.

Notes to Chapter 2

1. C. W. Hufeland, *The Art of Prolonging Life* (London, 1797), p. 73.
2. *Biological Rhythms in Psychiatry and Medicine* (Washington: National Institute of Mental Health, 1970), p. 83.
3. Ritchie R. Ward, *The Living Clocks* (New York: Alfred A. Knopf, 1971), p. 34.
4. J. N. Berrill, "Living Clocks," *Atlantic Monthly*, December 1963, p. 65.
5. Frank A. Brown, "Life's Mysterious Clocks," *Saturday Evening Post*, 24 December 1960, p. 19.
6. Frank Brown, *A Unified Theory for Biological Rhythms, Circadian Clocks* (Amsterdam, 1965).
7. Michel Gauquelin, *The Cosmic Clocks* (1967; reprint ed., New York: Avon Books, 1969), p. 126.
8. Berrill, "Living Clocks," p. 66.
9. Brown, "Life's Mysterious Clocks," p. 43.
10. Berrill, "Living Clocks," p. 66.
11. Ward, *The Living Clocks*, pp. 208-9.
12. Eric T. Pengelley and Sally J. Asmundson, "Annual Biological Clocks," *Scientific American*, April 1971, pp. 74-75.
13. Ward, *The Living Clocks*, pp. 234-37.
14. *University of Chicago Reports* 20, no. 2 (Fall 1970).
15. Ward, *The Living Clocks*, p. 295.
16. *Northrop Technical Digest* (1966); paper by Dr. Lindberg, 1968; interview with Dr. Lindberg, 1971.

Notes to Chapter 3

1. *Wall Street Journal*, 29 July 1970.
2. Fielding H. Garrison, *History of Medicine* (1929; reprint ed., Philadelphia: W. B. Saunders Co., 1960), p. 260.
3. *Biological Rhythms in Psychiatry and Medicine* (Washington: National Institute of Mental Health, 1970), pp. 8-9.
4. Ibid., pp. 2-3.
5. J. N. Mills, "Human Circadian Rhythms," *Physiological Reviews* 46 (January 1966): 155.
6. Canadian Press Service, January 1970.
7. *Biological Rhythms*, p. 10.
8. Andrew Hamilton, "The Mystery of the Biological Clocks," *Science Digest*, October 1964, p. 22.

9. *Science* 130 (December 1959): 1538.
10. *National Observer,* 28 December 1970.
11. *Science* 130 (December 1959): 24.
12. *Biological Rhythms,* p. 155.
13. *National Observer,* 28 December 1970.
14. *News Front,* May 1970, p. 45.
15. *Life,* 24 July 1970, p. 10.
16. Mills, "Human Circadian Rhythms," p. 128.
17. *Biological Rhythms,* p. 44.
18. Mills, "Human Circadian Rhythms," p. 136.
19. *Biological Rhythms,* p. 45.
20. Mills, "Human Circadian Rhythms," p. 143.
21. *Biological Rhythms,* p. 46.
22. Ibid., p. 48.
23. Ibid., p. 49.
24. *Biochemistry,* 8 July 1967.
25. *Biological Rhythms,* p. 63.
26. Ibid., p. 140.
27. Mills, "Human Circadian Rhythms," p. 129.
28. Erwin Bunning, *The Physiological Clock* (New York: Springer-Verlag, 1967), p. 3.
29. Ritchie R. Ward, *The Living Clocks* (New York: Alfred A. Knopf, 1971), p. 363.
30. *Biological Rhythms,* p. 123.
31. Ibid., p. 141.
32. Ibid., pp. 126-32.

Notes to Chapter 4

1. *Biological Rhythms in Psychiatry and Medicine* (Washington: National Institute of Mental Health, 1970), p. 15.
2. Dr. Paul Naitoh, Navy Medical Neuropsychiatric Research Unit, San Diego, California, writing in *Minneapolis Tribune,* 1 March 1971.
3. Dr. Naitoh, in *Minneapolis Tribune,* 15 March 1971.
4. Dr. Naitoh, in *Minneapolis Tribune,* 8 March 1971.
5. Ibid.
6. *Stanford Observer,* February 1971.
7. Dr. Naitoh, in *Minneapolis Tribune,* 22 March 1971.
8. *Biological Rhythms,* p. 17.

9. Ernest Hartmann, *The Biology of Dreaming* (Springfield, Ill.: Charles C. Thomas, 1967), pp. 51-73.
10. Dr. Naitoh, in *Minneapolis Tribune,* 15 March 1971.
11. *New Scientist,* 23 April 1970, pp. 170-72.
12. *News Front,* May 1970, p. 45.
13. R. Rubin, in *Montreal Star,* 10 October 1970.

Notes to Chapter 5

1. Maurice Zolotow, "Jet Upset," *American Weekly,* 25 August 1963, pp. 10-11.
2. *Time,* 17 December 1965, p. 66.
3. *Montreal Star,* 22 May 1971.
4. *Archives of Neurology* 22 (June 1970): 483-89.
5. *Montreal Star,* 22 May 1971.
6. *New York Times,* 15 September 1968.
7. *Science News* 100 (20 November 1971): 343.
8. *Biological Rhythms in Psychiatry and Medicine* (Washington: National Institute of Mental Health, 1970), p. 138.
9. Marc Richelle, "Biological Clocks," *Psychology Today,* May 1970, p. 35.
10. Lowell Thomas, "Keep an Eye on Your Internal Clock," *Reader's Digest,* August 1966, pp. 61-64.
11. *Biological Rhythms,* p. 139.
12. Ibid., p. 135.
13. Ibid., p. 137.
14. Ibid.
15. Ibid., p. 21.
16. Technical paper by Drs. Beers and Rummel, NASA, 1970.
17. Ronald Murton, "Behind the Face of the Biological Clock," *New Scientist and Science Journal,* 29 July 1971, pp. 248-50.
18. *Los Angeles Times,* 13 July 1971.

Notes to Chapter 6

1. *Chatelaine Magazine,* December 1968, p. 77.
2. *Nervenarzt* (1968).
3. *Chatelaine Magazine,* December 1968, p. 77.
4. *New York Times,* 3 May 1971.

5. *Chatelaine Magazine,* December 1968, p. 78.
6. *Biological Rhythms in Psychiatry and Medicine* (Washington: National Institute of Mental Health, 1970), p. 14.
7. Ibid., pp. 100-103.
8. All material following Note 7 is from Charles F. Stroebel, "The Importance of Biological Clocks in Mental Health," *National Institute of Mental Health Reports,* 1968, no. 2 (February), pp. 323-52.

Notes to Chapter 7

1. *World Book Encyclopedia,* 1966 ed., s.v. "Charles Lamb."
2. *Biological Rhythms in Psychiatry and Medicine* (Washington: National Institute of Mental Health, 1970), p. 8.
3. *Chatelaine Magazine,* December 1968, p. 77.
4. Joan Lynn Arehart, "The Search for Clues to the Rhythms of Life," *Science News* 100 (11 September 1971): 178-79.
5. *Biological Rhythms,* p. 109.
6. Fielding H. Garrison, *History of Medicine* (1929; reprint ed., Philadelphia: W. B. Saunders Co., 1960), p. 36.
7. *Biological Rhythms,* p. 110.
8. Ibid.
9. Ibid., p. 112.
10. J. N. Mills, "Human Circadian Rhythms," *Physiological Reviews* 46 (January 1966): 158-59.
11. *Biological Rhythms,* p. 60.
12. Ibid., p. 61.
13. Ibid., pp. 66-67.
14. Mills, "Human Circadian Rhythms," pp. 156-57.
15. Ritchie R. Ward, *The Living Clocks* (New York: Alfred A. Knopf, 1971), pp. 238-39.
16. J. N. Berrill, "Living Clocks," *Atlantic Monthly,* December 1963, pp. 65-68.
17. *Biological Rhythms,* pp. 67-68.
18. Ibid., pp. 112-13.
19. Ibid., p. 116.
20. Fred Kerner, *Stress and Your Heart* (New York: Hawthorn Books, 1961), pp. 158-59.
21. *Johns Hopkins Magazine,* Spring 1968, pp. 18-19.
22. *Biological Rhythms,* pp. 42-43.

Notes to Chapter 8

1. *Biological Rhythms in Psychiatry and Medicine* (Washington: National Institute of Mental Health, 1970), p. 75.
2. E. M. Steindler, "Nature's Built-in Clocks," *Today's Health,* November 1965, pp. 55-63.
3. World Book Science Service article in *Montreal Star,* 8 July 1970.
4. *Biological Rhythms,* pp. 73-74.
5. Ibid., p. 68.
6. Marc Richelle, "Biological Clocks," *Psychology Today,* May 1970, p. 58.
7. All foregoing material following the reference to Note 6 is from *Biological Rhythms,* pp. 71-82.

Notes to Chapter 9

1. *Los Angeles Times,* 3 October 1971 and 19 October 1971.
2. J. N. Berrill, "Living Clocks," *Atlantic Monthly,* December 1963, p. 69.
3. *Biological Rhythms in Psychiatry and Medicine* (Washington: National Institute of Mental Health, 1970), pp. 37-38.
4. Marc Richelle, "Biological Clocks," *Psychology Today,* May 1970, pp. 59-60.
5. John E. Gibson, "Science Explores the Secrets of Time," *Today's Health,* August 1960, pp. 50-51.
6. Richelle, "Biological Clocks," pp. 59-60.
7. Gay Gaer Luce, "Understanding Body Time in the 24-Hour City," *New York Magazine,* 1971, pp. 40-52.
8. *Biological Rhythms,* pp. 55-56.
9. Luce, "Understanding Body Time," pp. 40-52.
10. *Biological Rhythms,* p. 152.

Notes to Chapter 10

1. Proceedings of the National Academy of Sciences, reported in *Science News* 100 (2 October 1971): 226.
2. J. N. Mills, "Human Circadian Rhythms," *Physiological Reviews* 46 (January 1966): 164.
3. E. M. Steindler, "Nature's Built-in Clocks," *Today's Health,* November 1965, p. 52.

4. *Biological Rhythms in Psychiatry and Medicine* (Washington: National Institute of Mental Health, 1970), p. 81.
5. *World Book Encyclopedia,* 1966 ed., s.v. "Yoga."
6. *New Scientist and Science Journal,* 9 September 1971, pp. 560-61.
7. *Montreal Star,* 20 September 1971.
8. *Los Angeles Times,* 7 March 1972.
9. *Time,* 25 October 1971, p. 51.
10. *Encyclopaedia Britannica,* 1961 ed., s.v. "Stigmatization."
11. "Programmed Biological Obsolescence," *Johns Hopkins Magazine,* Spring 1968, pp. 18-19.
12. *Los Angeles Times,* 9 August 1970.

Select Bibliography

Bunning, Erwin. *The Physiological Clock.* New York: Springer-Verlag, 1967.

Cohen, John. *Psychological Time in Health and Disease.* Springfield, Ill.: Charles C. Thomas, 1967.

Fraisse, Paul. *The Psychology of Time.* New York: Harper & Row, 1963.

Fraser, Julius T., ed. *The Voices of Time.* New York: George Braziller, 1966.

Gauquelin, Michel. *The Cosmic Clocks.* Chicago: Henry Regnery Co., 1967. Reprint. New York: Avon Books, 1969.

Mills, J. N. "Human Circadian Rhythms." *Physiological Reviews* 46 (January 1966).

BIBLIOGRAPHY

National Institute of Mental Health. *Biological Rhythms in Psychiatry and Medicine.* Washington, 1970.

Piltz, Albert, and Van-Bever, Roger. *Time Without Clocks.* New York: Grosset & Dunlap, 1971.

Richter, Curt P. *Biological Clocks in Medicine and Psychiatry.* Springfield, Ill.: Charles C. Thomas, 1965.

U.S. Department of Health, Education and Welfare. *Mental Health Program Reports No. 2.* Washington: 1968.

Ward, Ritchie R. *The Living Clocks.* New York: Alfred A. Knopf, 1971.

Identifying Cycle of Strength, Endurance, and Courage

(Record periods of greatest strength and weakness by making a chart like the one on p. 187 and filling it in according to the general method described on p. 186. Such a chart will help in scheduling physically arduous activities.)

Identifying Cycle of Sensitivity, Intuition, and Love

(Record periods of greatest and least sensitivity by making a chart like the one on p. 187 and filling it in according to the general method described on p. 186. Such a chart will indicate the best times for pursuing interests of the heart.)

Identifying Cycle of Memory, Alertness, and Reasoning Power

(Record periods of greatest and least intellectual ability by making a chart like the one on p. 187 and filling it in according to the general method described on p. 186. Such a chart will indicate the best times for undertaking tasks requiring mental acuity.)

Identifying Daily Cycle of Sensory Keenness

(Over a period of several days, record the hours at which you are most sensitive to sights, sounds, smells, tastes, and sensations of touch. Such a record will reveal those times of day during which you will most enjoy such activities as watching a movie or play, listening to music, eating, and making love.)

Discovering an Ideal Schedule

(For several days on which you do not have to work or attend school, rule a pad of paper vertically into twenty-four hourly segments and check off the hours when you normally feel like eating, going to bed, getting up, exercising, etc. Finding ways to make your ordinary working day routine conform more closely to this ideal schedule can greatly enhance the sense of well-being.)